Building-Integrated Photovoltaic Systems (BIPVS)

Andrés Julián Aristizábal Cardona
Carlos Arturo Páez Chica
Daniel Hernán Ospina Barragán

Building-Integrated Photovoltaic Systems (BIPVS)

Performance and Modeling Under Outdoor Conditions

Andrés Julián Aristizábal Cardona
Engineering Department
Universidad de Bogota Jorge
Tadeo Lozano
Bogota, Colombia

Carlos Arturo Páez Chica
Engineering Department
Universidad de Bogota Jorge
Tadeo Lozano
Bogota, Colombia

Daniel Hernán Ospina Barragán
Engineering Department
Universidad de Bogota Jorge
Tadeo Lozano
Bogota, Colombia

ISBN 978-3-319-89122-4 ISBN 978-3-319-71931-3 (eBook)
https://doi.org/10.1007/978-3-319-71931-3

Softcover re-print of the Hardcover 1st edition 2018

Printed on acid-free paper

This Springer imprint is published by Springer Nature
The registered company is Springer International Publishing AG
The registered company address is: Gewerbestrasse 11, 6330 Cham, Switzerland

Contents

List of Figures

List of Tables

Introduction

Electricity generation located near the energy load is called distributed generation. The hardware is typically installed in the same area as the energy demand. There is a wide variety of technology that covers distributed generation, although their use is based on availability.

The present work reports on the design, installation, and start-up of the building integrated photovoltaic system (BIPVS) that is connected to the low-voltage grid of the Universidad de Bogotá Jorge Tadeo Lozano's Engineering Programs' Center for Research (CIPI). It also describes a mathematical analysis that models the behavior of a solar panel in the system. This book also presents the design and installation of a monitoring system for the performance of the solar plant that uses virtual instrumentation and analysis of the results of the photovoltaic (PV) generator and inverter.

The photovoltaic generator is made up of 24 solar panels, Trina Solar, of 250 W each and a Sunny Boy two-phase inverter of 5000 W. The results of the modeling stage indicate a relative error of 1.5% between the maximum power point reported by the solar panel manufacturer and the proposed mathematical analysis under a solar radiation level of 1000 W/m^2. The main results of the performance monitoring analysis of the BIPVS system show that over a period of two years, 16.194 kWh was generated and the harmonics, nominal voltage, and system frequency have been within the operating ranges suggested by the IEEE 929-2000 standard. Solar radiation has exceeded 4 kWh/m^2-day.

The book consists of 12 chapters, providing a comprehensive overview of the topic at hand.

Chapter 1 presents the current state of energy worldwide, with emphasis on its utilization in Colombia.

Chapter 2 corresponds to the conceptual framework of the work developed. It provides a global view on distributed energy generation and applies the concepts of BIPVS, with its respective form of operation and distributed generation (DG), among other concepts.

Chapter 3 provides the basics of the BIPVS sizing, power quality, monitoring system, and virtual instrumentation.

Chapter 4 reviews a BIPVS application on a 40 story building.

Chapter 5 focuses on the sizing and implementation of the BIPVS installed at the Universidad de Bogotá Jorge Tadeo Lozano as an object of research and development.

Chapter 6 describes solar cell characteristics and presents a method for calculating short-circuit spectral current density and quantum efficiency.

Chapter 7 corresponds to the modeling and characterization stage of the photovoltaic solar panel to determine the reliability of the implemented model's operation.

Chapter 8 covers detailed implementation of the monitoring system to carry out the acquisition of environmental and electrical parameters of the photovoltaic generation system. These parameters include all the physical elements of the system such as equipment, transducers, power lines, protections, acquisition cards, and PC.

Chapter 9 refers to the performance, behavior, and analysis of BIPVS.

Chapter 10 corresponds to stable state analysis of the power system incorporating a 6 kW photovoltaic generator to the low-voltage grid of the CIPI building.

Chapter 11 evaluates BIPVS power generation through the use of neural networks.

Chapter 12 refers to the analysis and economic viability of the photovoltaic system implemented at the UTADEO University through the RETScreen software.

Chapter 1
Energy's Current State

Due to finite reserves of fossil fuels, limited hydroelectric potential and increasing demand for electricity as a result of the industry and world population growth, specialists predict that in the medium or long term electricity will have to be generated from new sources in order to maintain a balance between supply and demand for electricity, at both the global and national levels. In this context, the Colombian authorities evaluated this situation and, through the Ministry of Mines and Energy, issued Law 1715 of 2014, which promotes the use of renewable energy in the country, including photovoltaic electricity generation.

Taking into account the huge solar potential that exists in many regions of the country and the renewable nature of solar energy, photovoltaic generation is an excellent alternative that the country has to add to the conventional generation and, in this way, achieve in the future the balance between supply and demand. In addition, it is worth noting that solar energy is the source that more technologies have allowed the human beings to develop, counting among them the hydropower, wind, thermal and photovoltaic energy as the main ones.

In turn, it is worth mentioning that Colombia recently agreed to (under the COP21 Paris forum) reduce the emission of gases that generate earth's overheating. An important contribution to the reduction of the emission of this type of gases could be made by reducing in the future the growth of the thermoelectric generation and incorporating in its place solar plants to the Interconnected System. The substitution of diesel plants in not interconnected remote areas by solar plants could also contribute to the fulfillment of the goals set in the COP21 Paris forum.

Figure 1.1 shows the current global Renewable Energy landscape and its contribution to global primary energy by 2016 [1]. Renewable energy provided 19.3% of the world's final energy consumption, including traditional biomass, large hydropower, and "new" renewables (small hydropower, modern biomass, wind, solar, geothermal, and biofuels).

Traditional biomass was used mainly for food cooking and space heating and accounted for 9.1%; the large hydropower plants represented 3.6% and their growth was modest, especially in developing countries. New renewables accounted for

A.J. Aristizábal Cardona et al., *Building-Integrated Photovoltaic Systems (BIPVS)*,
https://doi.org/10.1007/978-3-319-71931-3_1

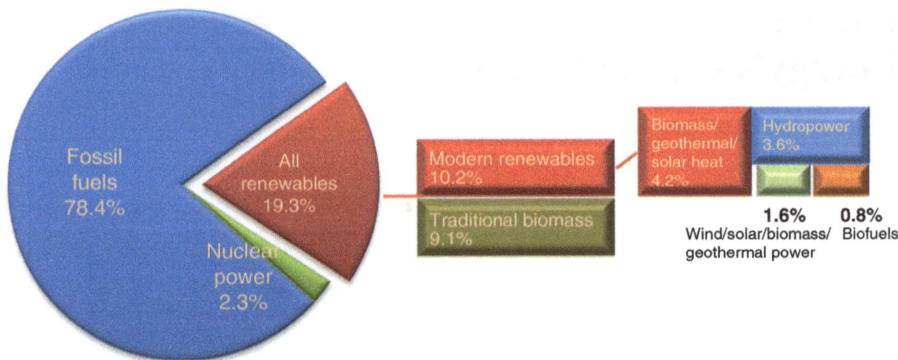

Fig. 1.1 Estimated Renewable Energy Share of Global Final Energy Consumption, 2015. Adapted from Renewable 2017 Global status report (p. 30), by REN21 (2016)

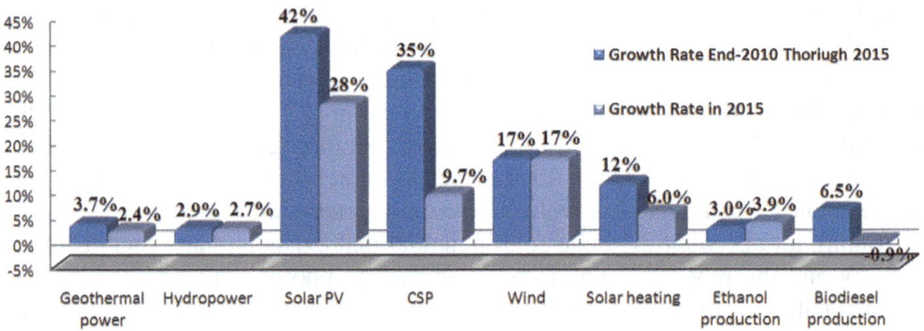

Fig. 1.2 Average Annual Growth Rates of Renewable Energy Capacity and Biofuels Production, End-2010 to End-2015. Adapted from Renewable 2016 Global status report (p. 29), by REN21 (2016)

1.6% and are growing very fast in developed countries. Renewable energy replaces fossil fuels in four different sectors: power generation, hot water and space heating, transportation fuel and rural energy.

Regarding the growth of renewable energies, Fig. 1.2 illustrates the behavior in this field by the end of 2015 [2].

The interconnected photovoltaic solar technology was the one with the highest growth, with an annual average of 42% for the 5-year period between 2010 and 2015. Other technologies that showed a high growth were concentrated solar power (35%), wind energy (17%), and solar energy for heating (12%). In conclusion, renewable energy growth is much greater than that of hydropower (2.9%), geothermal (3.7%), ethanol (3%), and biodiesel (−0.9%) for the same period of time.

Figure 1.3 presents the installed capacity of renewable power at the global level in 2016, discriminating the world regions that contribute most to this capacity [1]. In 2016, global renewable power capacity totaled 921 GW, of which, largest capacities are wind and solar photovoltaic. It is also observed that BRICS countries

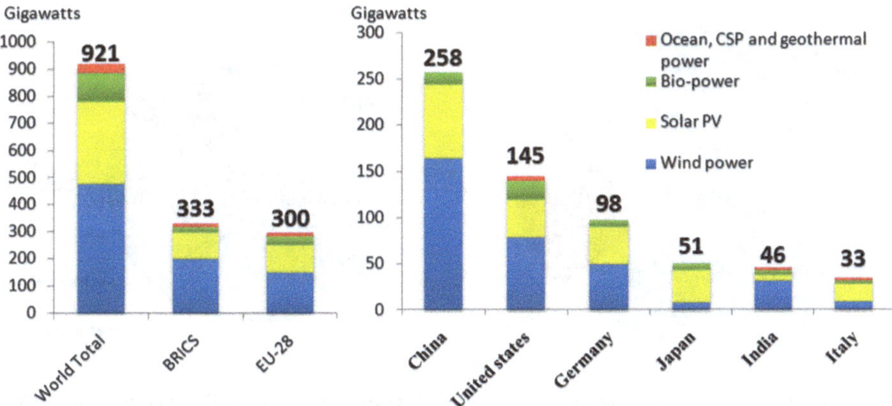

Fig. 1.3 Renewable Power Capacities in World, BRICS, EU-28 and top Six Countries, 2016. Adapted from Renewable 2017 Global status report (p. 34), by REN21 (2017)

Table 1.1 Renewable energy global potential, 2016

Technology	Technical potential scale (TW) h/year)
Photovoltaic	12,000–40,000
Wind	20,000–40,000
Ocean	2000–4000
Geothermal	4000–40,000
Biomass	8000–25,000

(Brazil, Russia, India, China, and South Africa) and the European Union contributed most to this capacity. BRICS countries contributed 333 GW, mostly from wind energy and solar photovoltaic, surpassing the European Union that contributed 300 GW. Finally, China was the first among the top six countries, with a total of 258GW.

Renewable energies have a great generation potential as shown in Table 1.1.

Figure 1.4 shows that 24.5% of total electricity generation comes from renewables, in greater proportion to hydraulic generation 16.6% and wind power 4.0%. Solar, geothermal, CSP, oceans and biomass totaled 3.9% [1].

Typical costs for generating electricity with renewables are shown in Table 1.2. Solar photovoltaic energy in buildings has prices ranging from 2200 to 7000 USD/kW installed, depending on the country where the project is located for installed capacities between 3 and 500 kW. Geothermal generation prices between $ 1900 and $ 5500/kW, and biomass combustion for plants between 1 and 200 MW has costs around $ 800–4500/kW.

Figure 1.5 shows investments in 2016 for renewable energy from developed and developing countries. The largest investments are made in photovoltaic solar energy, 57.5B USD, in developing countries and wind, 60.6B USD, in developed countries [1].

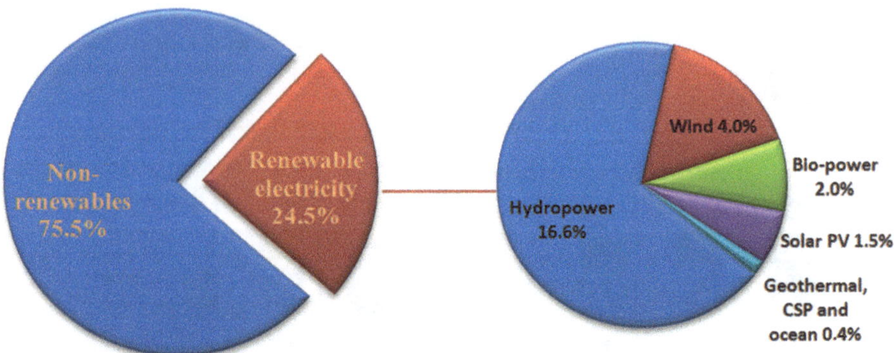

Fig. 1.4 Estimated Renewable Energy Share of Global Electricity Production, End–2016. Adapted from Renewable 2017 Global status report (p. 32), by REN21 (2017)

Figure 1.6 shows the global photovoltaic solar energy capacity, 2005–2015. In 2015 there is an addition of 50 GW compared to 2014, marking a growth due to greater investments in this technology [1].

The 10 countries that invested more in solar PV capacity by 2015 are shown in Fig. 1.7. China added 15.2 Gigawatts in capacity in 2015, Germany 1.5 GW, Japan 11 GW, USA 7.3 GW, Italy 0.3 GW, UK 3.7 GW, France 0.9 GW, Spain 0.1 GW, India 2 GW, and Australia 0.9 GW [1].

In Colombia, the most exploited renewable energy resources, with information recorded in the literature, have to do with technologies related to wind energy, energy from small hydroelectric plants, biomass energy (Biofuels specifically) and photovoltaic solar energy [3].

In terms of photovoltaic generation of power in Colombia there has been a moderate development. There are reports that show that up to 1990, about 1.8 MWp of photovoltaic power had been installed in the country in about 28,000 autonomous systems. The facilities were mostly carried out by the users themselves and about a third of these were used in telecommunication projects. A smaller part of these systems were used for road signaling and for meteorological stations [4].

In October 1992, a solar powered lighting system was installed in the "Community Health Center" in Mosoco, a Paez indigenous community in the mountainous region of Tierradentro. In 1995 a project in the Chocó carried out by the "World Health Organization", the "Pan American Health Organization" and by the Netherlands Government allowed to improve health services for rural communities by installing four different PV systems with approximately 700 Wp in each community. Two of the systems were used for rural health centers in each community, refrigeration and other applications; one supplied electricity to the home of the community doctor and the last one was used for a video room equipped with a TV and a VCR. Two inhabitants of each community were trained in the installation, maintenance and repair of the systems. Therefore, the project was characterized by the great participation of the community and was well accepted [5, 6].

Table 1.2 Costs of power generation. Adapted from Renewable 2016 Global status report, by REN21 (2016)

Technology	Typical characteristics	Capital costs USD/kW
Power generation		
Bio-power from solid biomass (including co-firing and organic MSW)	Plant size: 1–200 MW Conversion efficiency: 25–35% Capacity factor: 50–90%	800–4500 Co-fire: 200–800
Bio-power from gasification	Plant size: 1–40 MW Conversion efficiency: 30–40% Capacity factor: 40–80%	2050–5500
Bio-power from anaerobic digestion	Plant size: 1–20 MW Conversion efficiency: 25–40% Capacity factor: 50–90%	Biogas: 500–6500 Landfill gas: 1900–2200
Geothermal power	Plant size: 1–100 MW Capacity factor: 60–90%	Condensing flash: 1900–3800 Binary: 2250–5500
Hydropower: Grid-based	Plant size: 1–18,000+ MW Plant type: reservoir, run-of-river Capacity factor: 30–60%	Projects >300 MW: 1000–2250 Projects 20–300 MW: 750–2500 Projects <20 MW: 750–4000
Hydropower: Off-grid/rural	Plant size: 0.1–1000 kW Plant type: run-of-river, hydrokinetic, diurnal storage	1175–6000
Ocean power: Tidal range	Plant size: < 1 to > 250 MW Capacity factor: 23–29%	5290–5870
Solar PV: Rooftop	Peak capacity: 3–5 kW (residential); 100 kW (commercial); 500 kW (industrial) Capacity factor: 10–25% (fixed tilt)	Residential costs: 2200 (Germany); 3500–7000 (USA); 4260 (Japan); 2150 (China); 3380 (Australia); 2400–3000 (Italy) Commercial costs: 3800 (USA); 2900–3800 (Japan)
Solar PV: Ground-mounted utility-scale	Peak capacity: 2.5–250 MW Capacity factor: 10–25% (fixed tilt) Conversion efficiency: 10–30% (high end is CPV)	1200–1950 (typical global); as much as 3800 including Japan. Averages: 2000 (USA); 1710 (China); 1450 (Germany); 1510 (India)
Concentrating solar thermal power (CSP)	Types: parabolic trough, tower, dish Plant size: 50–250 MW (trough); 20–250 MW (tower); 10–100 MW (Fresnel) Capacity factor: 20–40% (no storage) 35–75% (with storage)	Trough, no storage: 4000–7300 (OECD) 3100–4050 (non-OECD) Trough, 6 h storage: 7100–9800 Tower: 5600 (USA, without storage) 9000 (USA, with storage)
Wind: Onshore	Turbine size: 1.5–3.5 MW Capacity factor: 25–40%	925–1470 (China and India) 1500–1950 (elsewhere)
Wind: Offshore	Turbine size: 1.5–7.5 MW Capacity factor: 35–45%	4500–5500

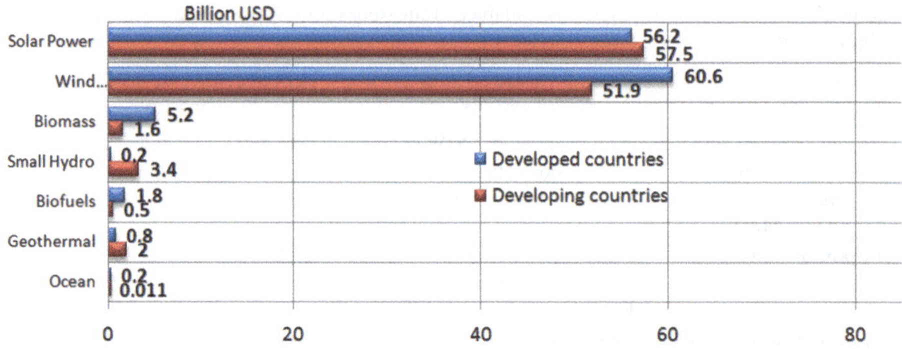

Fig. 1.5 Global new investment in renewable energy by technology, Developed and developing countries, 2016. Adapted from Renewable 2017 Global status report, by REN21 (2017)

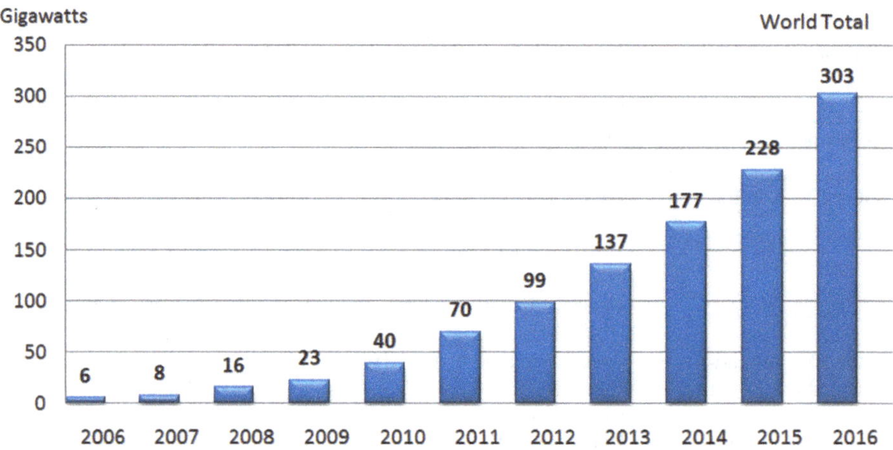

Fig. 1.6 Solar PV global capacity, 2006–2016. Adapted from Renewable 2017 Global status report, by REN21 (2017)

In July 2004 an offer to install 54 hybrid solar–diesel systems to provide electricity to enable remote community centers (Telecenters) to carry out voice and data transmissions including telephony, fax and Internet services allowed proposals to be submitted by Total Energie SNC, BP Solar, Shell Solar and Emerson Energy. In September 2004, the French photovoltaic company Total Energie SNC obtained the project, which was developed by its local subsidiary Tenesol Colombia in cooperation with the Ministry of Telecommunications and it could benefit 18,000 people by providing them Internet, telephony and fax communications [7].

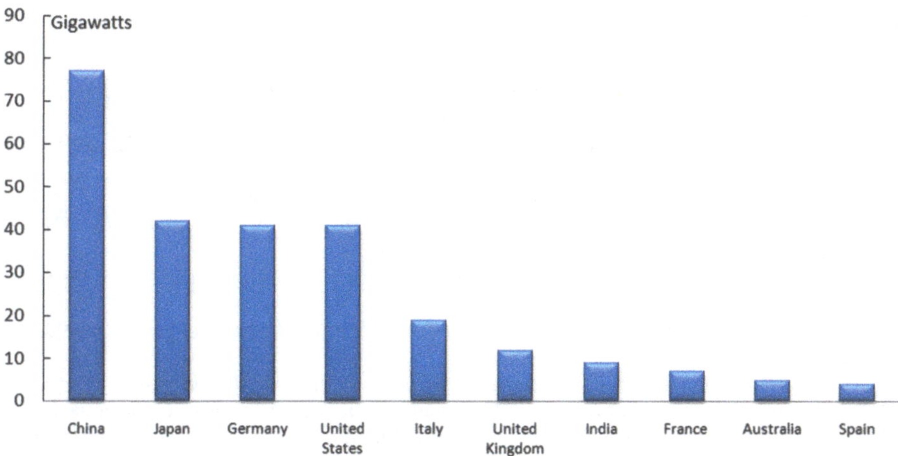

Fig. 1.7 Solar PV capacity, top 10 countries, 2016. Adapted from Renewable 2017 Global status report, by REN21 (2017)

In recent years there are several reports that inform on the installation of various photovoltaic systems for different applications. That is the case of the Universidad Pontifica Bolivariana in Medellin, which has a projected capacity of 70 kWp. The Universidad Jorge Tadeo Lozano has installed a 6 kW interconnected photovoltaic solar system in 2015 and plans to install another of the same capacity in 2017. The Universidad Autonoma de Occidente has installed 150 kW in solar panels. Other educational and commercial centers like Grupo Éxito, Jumbo among others have already made significant investments in photovoltaic solar energy in the country.

The term "Building Integrated Photovoltaic Systems (BIPVS)" refers to the concept of integrating photovoltaic elements in the building envelope, establishing a symbiotic relationship between architectural design, functional properties, and economic regeneration of energy conversion [8]. This type of installation has a photovoltaic generator and an inverter; therefore, the energy delivered to the grid by a BIPV system is the one that produces its generator minus the one that loses its inverter [9].

Colombia is currently experiencing an energy crisis caused by the country's reduced capacity to supply electricity due to the El Niño phenomenon, combined with the failure of some generators in the energy sector, causing a deficit in energy production. This type of crisis entails the regulation, integration and development of nonconventional renewable energy into the national energy system as a necessary means for sustainable economic development, greenhouse gas (GHG) reduction and security of energy supply; according to the general provisions, article 1 of Law 1715 of May 2014 [10].

References

1. Renewables 2017, Global Status Report. Eric Martinot, REN 21-Renewable Energy Policy Network for the 21st Century. (Worldwatch Institute, Washington, DC, 2017)
2. Renewables 2016, Global Status Report. Eric Martinot, REN 21-Renewable Energy Policy Network for the 21st Century. (Worldwatch Institute, Washington, DC, 2016)
3. "Energía Hídrica en Colombia", Retrieved from: www.aprotec.com.co
4. H. Rodríguez, S. Hurry, *Manual de Sistemas Fotovoltáicos para Electrificación Rural* (PNUD: OLADE: JUNAC, Bogota, 1995)
5. National Renewable Energy Laboratory (NREL): Energía Renovable para Centros de Salud Rurales, Sept. 1998, Golden, CO, USA
6. A. Jiménez, Energía Solar Internacional, Retrieved from: http://www.nrel.gov/docs/fy99osti/26224.pdf
7. PV Projects in Colombia: Tenesol, Retrieved from: Total Energie SNC (www.total-energie.fr)
8. E.C. Martín, Edificios Fotovoltaicos Conectados a la Red Eléctrica: Caracterización y Análisis, PhD Thesis, Universidad Politécnica de Madrid, Higher Technical School of Telecommunications Engineers. Madrid, 1998
9. R. Itusaka, Cálculo de la Energía Generada por un Sistema Fotovoltaico Conectado a red a 3800 msnm, Cusco–Peru, 2010
10. Law 1715 May 2014. Ministry of Mines and Energy

Chapter 2
Conceptual Framework

2.1 Power Generation with Photovoltaic Solar Energy

Solar energy is transformed directly into electricity by photovoltaic cells. This process is based on the application of the photovoltaic effect, which occurs when light hits on materials called semiconductors. Light is composed of photons, which are energetic particles. These photons are of different energies corresponding to the different wavelengths of the solar spectrum. When photons strike a photovoltaic cell, they can be reflected or absorbed, or can pass through it. Only the absorbed photons can generate electricity. When a photon is absorbed, the photon energy is transferred to an electron of an atom of the cell. With this new energy, the electron is able to escape its normal position associated with an atom to form part of a current in an electric circuit [1].

The equilibrium condition established in the semiconductor material by the PN juncture remains stable until the material N is exposed to light, the photon energy, which coincides with the value of the potential barrier, is absorbed by the material and destroys the bond of the valence electrons with the atom, then a chaotic movement of electrons within the material is caused. If an external charge is connected to the material, the electrons flow from the material and circulate through this external circuit thus releasing the energy absorbed from the photons [2].

The electricity produced by a photovoltaic panel is in direct current and its characteristic parameters (intensity and voltage) vary with the solar radiation that affects the cells and with the ambient temperature [3]. Electricity generated with photovoltaic solar energy can be transformed into alternating current, with the same characteristics as the electricity of the conventional grid, using inverters.

A.J. Aristizábal Cardona et al., *Building-Integrated Photovoltaic Systems (BIPVS)*,
https://doi.org/10.1007/978-3-319-71931-3_2

2.2 Types of PV Systems

2.2.1 Isolated Photovoltaic Systems

They are used in places with complicated access to the electricity grid and where it is easier and cheaper to install a photovoltaic system than to lay a line in the general grid. Its main objective is to satisfy all or part of the demand for electricity in these remote places. If it also relies on another power generation system such as a wind turbine or a generator set is called a mixed installation. Batteries are essential for storing energy when there is no sun, and to have energy throughout the whole day. In case of not owning batteries, the energy must be consumed instantaneously without possibility of storage. The elements that make up these isolated systems are: Photovoltaic Panel, Battery, and Regulator.

2.2.2 Grid-Connected Photovoltaic Systems

Photovoltaic systems connected to the grid can be of many different sizes and can range from small systems installed, for example on tiling or rooftops, to photovoltaic plants installed in large terrain (rural areas not used for other activities can be of benefit), or intermediate facilities such as those that use large surfaces of urban areas, parking lots, shopping centers, sports areas, etc.

The rooftop or large surface installations represent a clear example of some of the great advantages of photovoltaic energy. As an example, electricity generation can be produced in the same place where there is power consumption (which avoids costs, transportation losses and take full advantage of it, due to the high reliability of the grid [4]. on BIPVS, the photovoltaic generator is interconnected to the grid by means of an inverter. When the generation exceeds the consumption in a certain instant the system injects energy or extracts energy from it otherwise.

2.2.3 BIPVS Operability

The operation of the photovoltaic generation system interconnected to the low voltage grid is described by every equipment and devices that are part of the system.

- Photovoltaic modules: These are modules or photovoltaic cells that are designed to generate energy to supply the demand required by a load.
- Inverter: It is the equipment that transforms the energy received from the photovoltaic generator (in the form of direct current) and adapt it to the conditions required according to the type of loads, usually in alternating current and the following supply to the grid. The inverters are characterized mainly by the

input voltage, which must be adapted to the photovoltaic module, the maximum power it can provide and efficiency.

- General Distribution Board or Point of Common Coupling (PCC): A locker where the main power distribution bay is located and where the electrical system protection devices are housed to guarantee the personnel and installed equipment safety.
- Bidirectional Energy Meter: Measurement equipment that records the energy being consumed from the external grid and the energy delivered to the external interconnected system, delivered by the BIPVS.
- Electric load: Formed by all those appliances that demand power for its operation.
- External Electric grid: Energy system that is supplied by the grid operator to end users.

Figure 2.1 shows the configuration of the photovoltaic system connected to the grid.

2.2.4 Distributed Generation (DG)

In recent years, the electricity sector has shown attention in DG due to several factors such as: advances in small-scale generation technologies, liberalization of the electricity sector, and renewed ecological awareness [5]. The DG is the generation and storage of small-scale electrical energy, the closest to the load center, with the option of buying or selling power with the interconnected system [6].

Another cause that generates greater interest in DG is the contribution to sustainable development; since this is commonly associated with the production of clean energies [7] and contribution to energy security by improving system reliability levels [8].

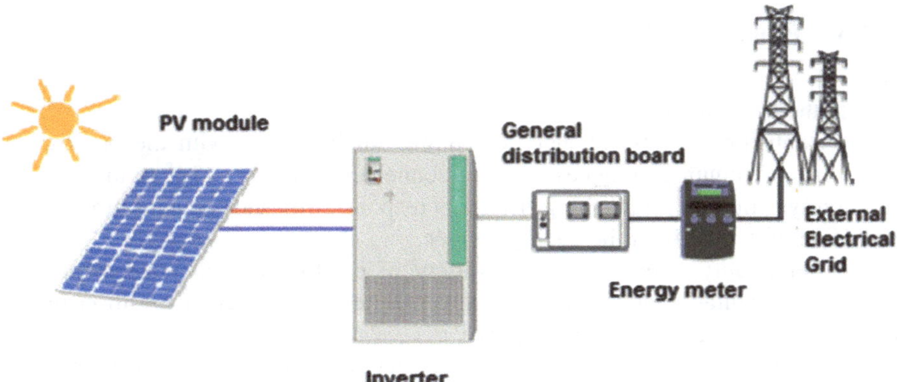

Fig. 2.1 BIPVS operation diagram

Distributed generation (DG) emerges as an important alternative for the provision of the power service, since it can increase the reliability and security of supply in the medium, short, and long term.

For the specific case of Colombia it is considered that the DG is promoted in its national development plan, because its privileged location it has several resources to implement clean technologies (biomass, solar, wind, and geothermal power) and ensure an efficient energy supply in Non-Interconnected Zones (NIZ), a case that is already being presented on a low scale [9].

Many advantages and disadvantages of distributed generation are directly related to the operation, stability, consistency, and reliability of the power system, in which new businesses have been created where both centralized and distributed generation resources can be integrated for obtaining higher incomes with lower prices [10].

Among the advantages of DG systems there are the following technical aspects, according to [11, 12]: reduction of losses in transmission lines, low-cost sources to meet demands during peak price periods, improvements in the power quality (Voltage waveform, frequency, voltage stability, reactive power supply, and power factor correction), high reliability sources for systems, sensitive users that cannot interrupt their power supply, reduction in GHG emissions (Renewable technologies), System backup generation in case of emergency, avoids large investments in transmission and distribution grids.

On the other hand, the negative impacts that are associated to the DG systems are [13, 14]: effects on the operation and maintenance of the distribution grid, voltage regulation in a continuous and transient state, decrease on the effectiveness of the distribution system protection.

The study of the DG impacts on the power quality highlights as a disadvantage the appearance of new problems of harmonics, flicker, transient surges, ferroresonance, power factor, deterioration of voltage profiles, decrease of system damping, and lines overloaded, among others [15].

2.2.5 Power Quality (PQ)

Due to the need to provide a better service to the end user, the grid operators have been establishing criteria of electrical power quality (PQ), with the purpose of providing minimum guarantees to users connected to the regional or local transmission system; thus establishing criteria of responsibility and compensation for the provision with quality of the energy service.

"Power quality refers to a wide variety of electromagnetic phenomena that characterize voltage and current at a given time and at a given position in the power system" IEEE 1159 (1995) [16].

Power quality in Colombia must meet specific regulations both for the grid Operator and the end Users that are connected to the electricity grid. The importance of standardization also lies in the fact that it serves as a reference point for studies related to PQ to be analyzed from the same perspective. The following are the most relevant and demanding standards in Colombia:

- IEC 61000-4-7 Standard: Electromagnetic Compatibility.
- IEEE Standard 519 (1992): The purpose of this standard is to recommend the practices and requirements of harmonic control in power systems due to nonlinear loads.
- IEEE Standard 1159 (1995): This standard deals with the monitoring of PQ for single-phase and poly phase AC systems.
- IEEE 1250 Standard: Describes the momentary voltage disturbances that occur in distribution systems, their identification, effect, and ways to meet the voltage requirements on sensitive equipment.
- IEEE 929-2000 standard: This standard explicitly defines power quality parameters for PV systems interconnected in parallel to the power grid and establishes the corresponding limits. The analysis parameters are: Total percentage of harmonic distortion and harmonic components, frequency, system's voltage, flicker, and power factor.
- NTC 5000: Power quality.
- NTC 1340: Nominal voltage and frequency in power systems in service networks.
- NTC 152: Limits and evaluation methodology in points of common coupling PCC.
- RETIE: Technical Regulation of Electrical Installations.

Power quality has been taken into account from the 80s, but in Colombia the topic took attention from the 90s. The power quality is related to the events or happenings that occur to the load due to possible problems in the electric power supply [17].

For the definition of power quality, three proposals can be considered, respectively:

1. Proposal by the Dugan, McGranaghan, and Beaty Engineers [18]

   "Any power problem manifested in the deviation of voltage, current or frequency resulting in failure or malfunction of a user's equipment."
2. Proposal by IEC (1000-2-2/4) and CENELEC (50160) [19]

   "The quality of energy or power is a physical characteristic of the electricity supply, which is delivered under normal operating conditions without disturbing the customer's processes."
3. Proposal by IEEE 1159 (1995) [20]

   "The quality of electrical energy refers to a wide variety of electromagnetic phenomena that characterize voltage and current at a given time and at a given position in the power system."

Power quality can be quantified technically through the following parameters:

(a) *Waveform*. Obeys to a pure sinusoidal form, without having permanent disturbances and harmonics.

Due to it has been observed that the presence of nonlinear loads causes harmonic distortions in power systems, in Colombia it was necessary a regulation to control the presence of these loads, so that normal operation would not be affected by their presence. The Colombian regulation establishes that both the grid operator and the Users (Clients) connected to its grid must comply with the IEEE 519 (1992) standard or the one it modifies or replaces.

In the matter of harmonics, both the grid operator and the User or Customer must comply with certain limits of distortion, both for current and voltage, established by the IEEE 519 standard.

(b) *Frequency*. It should be kept within a range established by the regulations, trying to be as constant as possible. The nominal frequency of the Colombian Interconnected System is 60 Hz and its range of operation variation is between 59.8 and 60.2 Hz under normal operating conditions. The grid operator and users should note that in emergency states, failure, energy deficits and reset periods the frequency can range from 57.5 to 63.0 Hz for a period of 15 s.

(c) *Symmetry*. The three-phase system must be as symmetrical as possible, in its alternating form, with 120° offsets.

(d) *Stability*. The effective value of the voltage must be kept as stable as possible. It is established that the voltages at steady state at 60 Hz and their permissible variations are those determined by the national standard NTC 1340 or the one that modifies or substitutes it. This standard establishes four voltage levels:

- Low voltage, for nominal system voltages less than or equal to 1000 V. Under normal system conditions it is recommended that the voltage at the supply terminals does not differ from the rated voltage by +5% and −10%.
- Medium voltage, for nominal voltages greater than 1000 V and less than or equal to 60,000 V. Under normal system conditions it is recommended that the voltage at the supply terminals does not differ from the rated voltage by +5% and −10%.
- High voltage, for nominal voltages greater than 60,000 V and less than or equal to 230,000 V;
- Extra high voltage, for values greater than 230,000 V. In these cases no ranges are specified but technical tables are given.

(e) *Rigidity*. The bars of the transmission system must be sufficiently rigid so that the voltage is not affected by possible variations due to the change in the state of the load. According to the electrical robustness at the connection points, or more specifically to the system's strength (which is directly related to the short-circuit levels), the system will be able to accept variations or changes in the load state without the disturbance or deterioration.

(f) *Reliability*. The reliability or availability of the system must be such that the customer is not affected by possible cuts. There are two indicators of service reliability. The first is an indicator of equivalent duration of service interruptions, and the second is the indicator of the equivalent frequency of service interruptions.

Table 2.1 Percentage of harmonic components

Odd harmonic	Distortion limit
3–9	<4.0%
11–15	<2.0%
17–21	<1.5%
23–33	<0.6%
>33	<0.3%

(g) *Voltage Flicker*. It is established that the flicker is a visible voltage disturbance between 15 and 30 Hz that should not appear in interconnected *systems*.

(h) *Waveform Distortion*. The output of the photovoltaic system must have low levels of harmonic distortion to ensure that there are no adverse effects on other equipment connected to the electrical grid according to the IEEE 519-1992 standard used to define acceptable levels of distortion for BIPVS. The total harmonic distortion must be less than 5% of the fundamental frequency in relation to the inverter output and each individual harmonic must be limited to the percentages listed in the following Table 2.1.

(i) *Power* Factor. The system must operate at a power factor greater than 0.85 leading or lagging.

(j) *Power of* Distortion. It is defined as voltamperes of distortion, which correspond to the product of voltage and current of components with different frequency.

(k) *Apparent* Power, *Reactive Power, Active Power*. These powers represent the description of the system electrical signals taking into account that they are affected by the content of harmonics originated by the equipment of conversion from direct current to alternating current.

References

1. A. Falk, C. Durschner, K.H. Remmers, *Photovoltaics for Professionals, Solar Electric Systems, Marketting, Design and Installation* (Solarpraxis AG, Berlin., Earthscan, London, 2007)
2. R.M.G. Castro, *Introdução à Energia Fotovoltaica* (Universidade Tecnica de Lisboa, Instituto Superior Técnico, Maio, 2008)
3. R. Khezzar, M. Zereg, A. Khezzar, *Comparative Study of Mathematical Methods for Parameters Calculation of Current-Voltage Characteristic of Photovoltaic Modules*, in Proceedings of International conference on Electrical and Electronics Engineering, (2009), pp. 24–28
4. T. Markvart, *Solar Electricity* (John Wiley and Sons, Inc., NY, 2000)
5. S. Stoft, *Power System Economics: Designing Markets for Electricity* (IEEE Wiley Interscience, Hoboken, 2002)
6. T. Ackermann, G. Andersson, L. Soder, Distributed generation: a definition. Electr. Power Syst. Res. **71**, 119–128 (2004)
7. International Energy Agency, *Contribution of Renewables to Energy Security* (IEA, Paris, 2007)
8. International Energy Agency, *Security of Supply in Electricity Markets: Evidence and Policy Issues* (IEA, Paris, 2002)

9. S.J. Murillo Joaquin, P. Roldán Carlos, *Model of Application of Distributed Generation in Colombia Rural Zones*. Transmission and Distribution Conference and Exposition (T&D), 2012 IEEE PES
10. C.I. Buriticá-Arboleda, C. Álvarez-Bel, *Decentralized Energy: Key to Improve the Electric Supply Security*. Conference on Innovative Smart Grid Technologies (ISGT Latin America), 2011 IEEE PES
11. D. Cristhian, M. Eduardo Felipe, R. de T., María Teresa, *Análisis de prospectiva de la generación distribuida (gd) en el sector eléctrico colombiano*. Revista de Ingeniería, n. 19, p. 81–89, may 2014
12. Y. Garzón, F. Tunarosa, *Smart Grids and Distributed Generation in Colombia*. Revista Vínculos, Universidad Distrital, n.10 (2), p. 303–309 (2013)
13. J. C. Gómez, Senior Member IEEE, *Generación Distribuida: Impacto en la Calidad de Potencia y en las Protecciones* (2008)
14. H. Torres, *Impacto en la Estabilidad de un Sistema de Potencia al Integrar Generación Distribuida* (Universidad Tecnológica de Pereira, pereira, 2008)
15. S. Carvajal, J. Marin, *The Impact of Distributed Generation on the Colombian Electrical Power System: A Dynamic-System Approach*. Tecnura n. 35 (17), p. 77–89 (2013)
16. IEEE Recommended Practice For Monitoring Electric Power Quality. IEEE Std 1159–1995
17. J. de Leonardo, C. Cardona, W.H. Zapata, J.H. Velásquez, R.H. Ortiz, *Quality of Electrical Power in a Distribution System* (Medellln, Colombia, 1999)
18. R.C. Dugan, M.F. McGranaghan, H. Wayne Beaty, *Electrical Power Systems Quality* (McGraw-Hill, New York, 1996)
19. Design of Reliable Industrial and Commercial Power Systems. CENELEC 493–1997
20. IEEE Recommended Practice for Monitoring Power Quality. IEEE Std 1159–1995

Chapter 3
BIPVS Basics for Design, Sizing, Monitoring, and Power Quality Measurement and Assessment

This chapter describes basic aspects of the design and sizing of interconnected photovoltaic systems (hereinafter referred to as BIPVS) and further details of the development of a high technology system implemented using the concept of virtual instrumentation to monitor and assess the technical performance and the power quality generated by BIPVS to the local power grid.

A Photovoltaic System Interconnected to the Electricity Grid is named after the fact that it has a source of photovoltaic electricity generation interconnected with the national system or the local grid. At present, interconnected photovoltaic systems are being used as a complement to conventional generation in many countries [1, 2]. Applications range from small-scale generation in remote areas in the past 25 years to the installation of centralized stations with large-scale photovoltaic generation capacity. Recently, an important stage of photovoltaic systems development was achieved, since this technology has transitioned from applications in remote areas to residential use at the urban level, where plants with small generation capacity are installed on the roofs of buildings or residences.

This latest application gave origin to the BIPVS (Building Integrated Photovoltaic Systems) concept that has been the focus of international attention in recent years. A very large number of residential PV systems have been installed internationally, in such a way that the generation with this type of PV systems now exceeds that generated with the others. Figure 3.1 schematically shows typical configurations of BIPVS for centralized and embedded operation (installed in residences) respectively.

In BIPVS, the photovoltaic generator is interconnected to the grid via an inverter. The inverters used to interconnect the DC output of the photovoltaic system with the public utility grid include an electronic paralleling control system that synchronizes the signal generated by the BIPVS with grid one, even in cases where several plants are connected in parallel. The inverter additionally includes a maximum power point tracking circuit that allows the PV generator to always transfer the maximum possible power, regardless of the type of load being connected to it. The efficiency of the inverter depends on the current supplied to

A.J. Aristizábal Cardona et al., *Building-Integrated Photovoltaic Systems (BIPVS)*,
https://doi.org/10.1007/978-3-319-71931-3_3

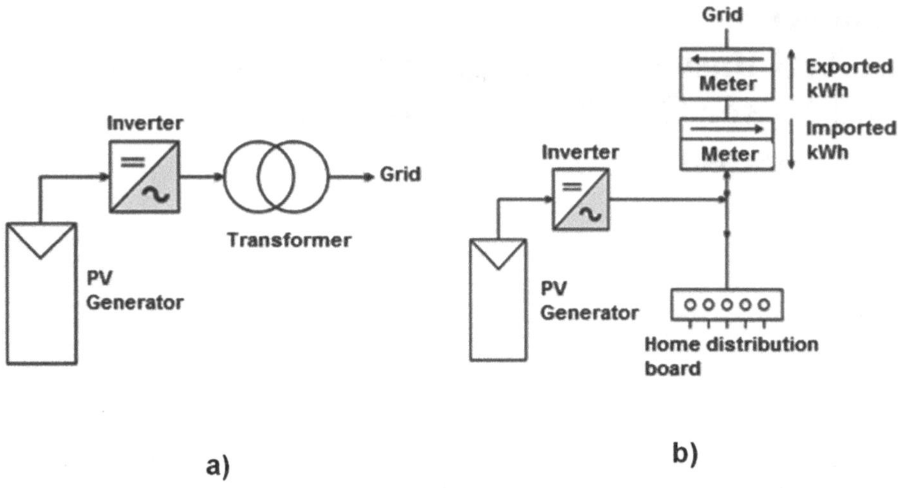

Fig. 3.1 BIPVS configuration for operation (**a**) centralized and (**b**) embedded in residential areas

the load, being maximum under operating conditions such that the power supplied to the load equals the inverter's nominal power. Efficiency can reach 95%; however, if it supplies less power than the nominal, its efficiency can be as high as 75%.

In centralized BIPVS, the AC voltage at the output of the inverter is raised through transformers to high or medium voltage levels, and subsequently, the electric fluid is brought to the user through the distribution grid. The pioneering work of Dan Shugar of Pacific Gas and Electric (PG & E) has shown that the use of centralized PV generation can be an effective solution to the problems of overloading and power quality at critical points in the interconnection grid [3, 4]. This is most strongly applied in developing countries, since in these countries the demand for electricity often exceeds generation capacity, especially in times of intense summer.

In embedded systems, or BIPVS systems, the arrangement of PV modules is placed on the roofs of buildings and these are interconnected to the local distribution grid. These systems allow taking electricity from the grid when the demand exceeds the PV generation and supply power to the grid when the PV generation exceeds the demand.

Recently, PV generation through BIPVS has had an explosive growth as a result of government subsidy policies for the residential use of photovoltaic systems. As a consequence of this, the photovoltaic industry has grown in the last 5 years at a rate of 30% [5, 6].

3.1 BIPVS Basics for Design and Sizing

BIPVS constitute the application of photovoltaic solar energy that has expanded the most in recent years. These systems have been integrating massively in the complex urban of numerous localities in many countries, and everything seems to indicate that this tendency will remain in the future.

Figure 3.2 shows a schematic form of the block diagram of a typical BIPVS. The design of this type of systems is done taking into account the following functional blocks:

PV generator, consisting of an array of PV modules interconnected in series and in parallel, with a configuration depending on the type of load and the characteristics of the inverter used. PV generator for BIPVS is installed in the roof of houses and the distribution and control panel is inside the house, turning it into a source of distributed or embedded generation.

The inverter, the power generated by the PV system is used primarily to supply electric fluid to the load and the surplus is exported to the grid. At night hours and on days of low solar radiation, the electric fluid consumed by the load is supplied by the grid. The inverter also incorporates some control functions that optimize the performance of the BIPVS (monitoring of the maximum power point of the photovoltaic generator, connection or disconnection from the grid depending on the conditions of it and the radiation incident on the generator and energy measurement, etc.).

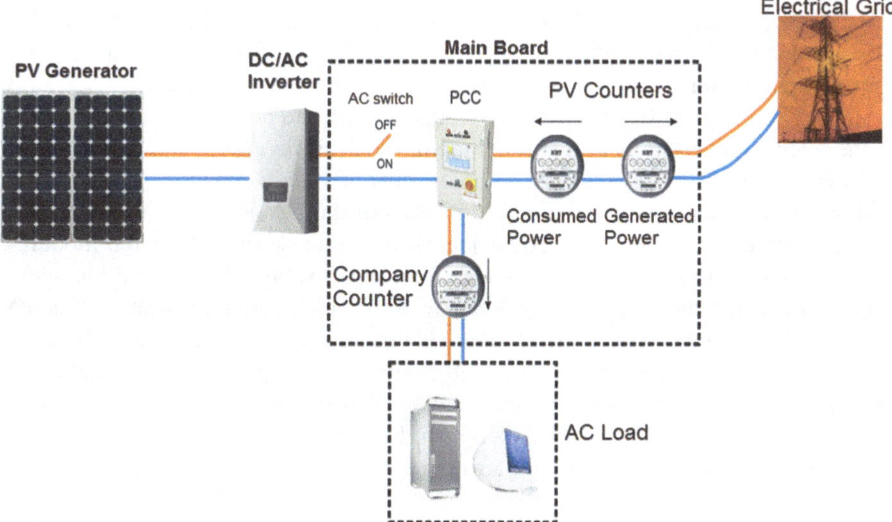

Fig. 3.2 Schematic of a BIPVS showing its five functional blocks

The technology of these devices has evolved significantly [7–9], but there are still limitations in their functioning [10–12]. The majority of the inverters used in BIPVS are self-switching and switching frequencies between 2 and 20 kHz, and nominal power between 1 and 50 kVA. In recent years, however, a new type of inverter, based on the switching of MOSFET transistors and with a nominal power of a few hundred W, has been designed to be installed directly on the back of the module [13–15]. Its main advantage lies in the simplicity of the installation, and its main disadvantage in the difficulty of the control that involves the installation of a large number of inverters in different places [16].

The distribution board, which includes meters to determine the exported and imported energy of the grid and the energy supplied to the load and protections to ensure the safety of the BIPVS itself and the safety of the electric grid and the PV interconnection system with the grid named point of common coupling (PCC). The load and the local grid are also part of the block.

3.1.1 Fundamentals for BIPVS Sizing

To size a BIPVS it is necessary to make a balance between the annual average of electric energy consumed by the load and the annual average of the electric energy generated by the BIPVS. The annual average of energy consumed is relatively easy to determine, as power companies perform statistics of this type and have this information available. On the other hand, the annual average of photovoltaic generated electricity is quite complicated to calculate with precision, since this requires the annual average of global solar radiation at the installation site, which is difficult to determine precisely, since it depends on climatic, geographic, and astronomical factors. Additionally, the calculation of the average annual PV energy is strongly influenced by loss factors associated with the power and characteristics of the selected inverter, as well as other factors such as: level of shading, orientation and inclination of the array of PV modules, dust and maintenance.

For the complete BIPVS, the measurements of the I–V and P–V characteristics are performed following a procedure identical to that described for the modules. However, these measurements are made under direct solar radiation and not with a solar simulator, as in the case of modules. In addition, due to high current of the PV generator, it is convenient to avoid methods that imply power dissipation by Joule effect [17, 18] and instead, it is recommended to use capacitive loads that allow measuring the I–V curve without large power dissipation.

A realistic approach to BIPVS sizing can be achieved using a model where the energy balance is realized using the principle of conservation of energy,

$$E_{FV} + E_{ER} = E_{VR} + E_{C} \tag{3.1}$$

Where: E_{FV} is the energy generated by the photovoltaic system, E_C the energy consumed by the building load, E_{ER} is the energy extracted from the electricity grid to supply consumption and E_{VR} the energy exported to the grid, surplus from consumption.

Equation (3.1) is applicable to any interval of time; however it will henceforth correspond to an annual period of time.

An expression equivalent to eq. (3.1), but normalized with respect to the EC variable, offers the advantage of independent analysis of the building specific consumption:

$$\frac{E_{FV}}{E_C} + \frac{E_{ER}}{E_C} = \frac{E_{VR}}{E_C} + 1 \tag{3.2}$$

Equation (3.2) can be written as a function of dimensionless variables, x and y:

$$x = \frac{E_{FV}}{E_C} \tag{3.3}$$

$$y(x) = 1 - \frac{E_{ER}(x)}{E_C} \tag{3.4}$$

So,

$$y = x - \frac{E_{VR}(x)}{E_C} \tag{3.5}$$

The variables x and y represent the relative capacity of the photovoltaic system (how much it is capable of producing, in relation to consumption) and the direct coverage of consumption (how much of the consumed comes from the photovoltaic system, therefore not being exchanged with the grid).

The variable x can be determined by calculating the energy produced by the photovoltaic system (PVS) knowing both the characteristics of the PV generator and the inverter as well as of the solar radiation and associated loss factors for geographic and astronomical effects. The variable y is determined by a study to the general behavior of BIPVS.

The loss factors affecting the energy produced by the PV generator are as follows:

1. Irradiation factor (IF)

   It is defined as the percentage of solar radiation incident on a PV generator installed with orientation and inclination (α, β) with respect to the corresponding one for optimal orientation and inclination (α=0°, $\beta_{opt.}$). Here α (azimuth angle) is the angle between the projection on the horizontal plane from the normal to the surface of the module and the meridian of the place and β is the angle of inclination that forms the surface of the module with the horizontal plane. The optimum tilt angle $\beta_{opt.}$ can be calculated by knowing the latitude ϕ of the installation location of the PV generator with the expression:

$$\beta_{\text{opt.}} = 3.7 + 0.69 * \phi \tag{3.6}$$

The FI Radiation Factor for given orientation and inclination can be determined using the expression:

$$\begin{gathered} \text{FI} = 1 - \left(1.2 * 10^{-4} \left(\beta - \beta_{\text{opt.}}\right)^2 + 3.5 * 10^{-5} \alpha^2\right) \\ \text{for}\, 15^\circ < \beta < 90^\circ \end{gathered} \tag{3.7}$$

$$\begin{gathered} \text{FI} = 1 - \left(1.2 * 10^{-4} \left(\beta - \beta_{\text{opt.}}\right)^2\right) \\ \text{for}\, \beta \leq 15^\circ \end{gathered} \tag{3.8}$$

2. Shading factor (SF)
   It is defined as the percentage of irradiation incident on the generator in the case of total absence of shadows. Shading losses are given by (1—FS). For PV generators with no obstacles on them and located on or near the equator, the approximate shadow factor is 8% i.e., SF = 0.92.
3. Incidence angle modifier factor (IAMF)
   This factor determines the additional losses of solar radiation that occur at sites deviated from the ideal position. This is determined by the following expression [19]:

$$\text{IAMF} = m_1 * \left(\beta - \beta_{\text{opt.}}\right)^2 + m_2 * \left(\beta - \beta_{\text{opt.}}\right) + m_3 \tag{3.9}$$

where:

$$m_j = m_{j1} * |\alpha|^2 + m_{j2} * |\alpha| + m_{j3}; \quad j = 1, 2, 3. \tag{3.10}$$

4. Global system performance factor (GSPR)
   This parameter determines the losses associated with the energy efficiency of the selected inverter. It can be calculated by the expression [19]

$$\text{GSPR} = \text{TF} * \eta_{\text{EI}} \tag{3.11}$$

Where η_{EI} is the energy efficiency of the inverter and TF is the temperature factor that indicates latitude-related temperature losses. These factors can be determined through expressions (3.12) and (3.13) [20].

$$\eta_{\text{EI}} = \frac{P_{\text{output}}}{P_{\text{input}}} = \frac{P_{\text{output}}}{P_{\text{output}} + \text{losses}} = \frac{p_0}{p_0 + \left(k_0 + k_1 \cdot p_0 + k_2 \cdot p_0^2\right)} \tag{3.12}$$

Where $P_0 = P_{\text{output}}/P_{\text{maxoutput}}$, is the output power normalized with respect to its maximum value, and

$$\mathrm{TF} = 1 - 0.065 * \sqrt{1 - \left(\frac{\phi - 25^{\circ}}{30}\right)} \quad (3.13)$$

Where

- k_0 represents loss of self-consumption (losses in the output transformer, control and operating devices, meters and indicators, etc.). It affects efficiency especially when the inverter operates at low load factor levels. A good inverter is characterized by losses of self-consumption below 1% [21].
- k_1 represents linearly dependent losses of operating power (diodes, switching devices, etc.).
- k_2 represents the losses that depend quadratically on the operating power (cables, coils, resistors, etc.).

5. Inverter sizing factor (FSI)
The desirability of oversizing the photovoltaic generator with respect to the inverter has been frequently revealed [22–24]. This loss can be quantified by a dimensionless parameter called

$$\mathrm{FSI} = \frac{P_{\mathrm{max,\,Inverter}}}{P_{\mathrm{nominal,\,Generator}}} = \frac{P_{\mathrm{max,I}}}{P_{\mathrm{nom,G}}} \quad (3.14)$$

3.2 Monitoring System Development Fundamentals

A monitoring application includes data acquisition, storage, and subsequent analysis. Although the type of data and its analysis vary, the tools needed to develop a monitoring system are very similar.

Monitoring systems have a set of basic characteristics. First, it is required to record the data in a storage system. Second, the data must be visible during the acquisition process (live) and after the acquisition. Third, it is needed to schedule alarms or events for the data. Fourth, implementing different types of data security should be an easy process. Fifth, grid operation must always be transparent to the user.

Viewing live data is required to monitor the status of a running system. The display of the data after it has been stored is known as historical display. These data are used for post-acquisition analysis and reporting.

The software used for the development of a monitoring system must have the tools and capacities that allow carrying out the said processes and in addition to functionally integrate the sensors, transducers and the hardware of conditioning and acquisition of data. The software must be flexible and have amenities to scale the system when it can be necessary to grow its complexity.

PC-based data acquisition systems basically have the configuration shown in Fig. 3.3.

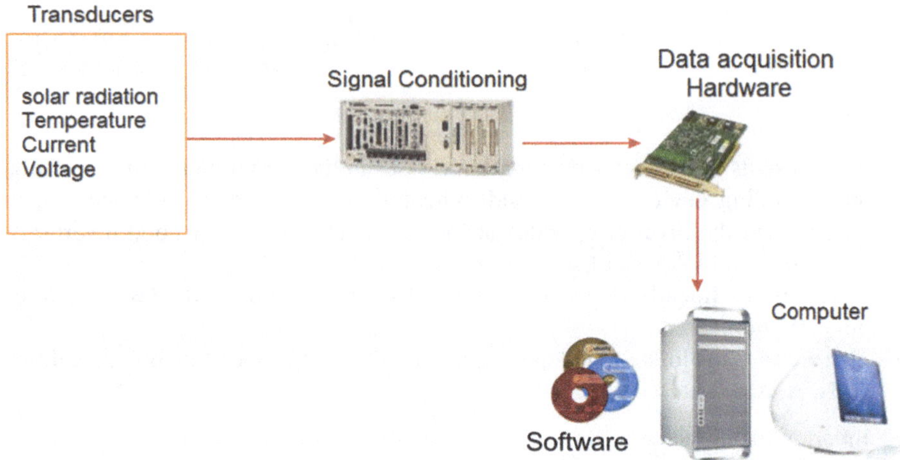

Fig. 3.3 Typical PC-based data acquisition system

3.2.1 Sensors and Signal Conditioning

The function of the sensors and transducers is to generate an electrical signal in response to the change of a physical parameter that is measured through the data acquisition hardware (DAH). These electrical signals must be conditioned for the input range of the DAH device. Signal conditioning devices basically perform attenuation, amplification, isolation and signal filtering to improve accuracy and resolution. Attenuation is necessary when the voltages to be digitized are higher than the input ranges of the digitizer, while filtering eliminates noise within a certain frequency range.

Since signal measurement generally includes 50 or 60 Hz noise from power lines or machines, most conditioners include low-pass filters specifically designed to provide maximum 50–60 Hz noise rejection. Another common use of filters is to prevent the signal "aliasing" (a phenomenon that arises when a signal is sampled very slowly). The Nyquist Theorem states that a signal must be sampled at least twice the highest frequency component of the signal to accurately reconstruct the waveform. To prevent a signal from being distorted by using a sampling rate lower than that of the Nyquist Theorem, a low pass filter can be used.

The lack of an adequate grounding system induces noise in the measured signals and exposes the acquisition equipment to damage. To prevent these problems, signal conditioning devices with insulation are used. These devices transfer the signals from the source to the measuring device without a physical connection using optical or capacitive coupling techniques. In addition to avoiding grounding, high-voltage insulation blocks avoid common-mode voltages and thus protect operators and measuring equipment.

The monitoring system developed in this work includes amenities to measure and monitor the current and voltage signals that give information about the power quality and technical performance of BIPVS, as well as solar radiation and ambient temperature. For this purpose a pyranometer and a thermistor were used as sensors of solar radiation and temperature, respectively. To measure the dc and ac current signals, current clamps were used as transducers.

(a) Global solar radiation sensor: pyranometer

Pyranometers are devices that have traditionally been used to measure global and diffuse solar radiation [25]. Among the parameters that characterize its performance are [26]: cosine response, detectivity, spectral selectivity, response time, response to the influence of external phenomena such as changes in temperature, etc. Based on these parameters the pyranometers are classified as first, second or third class [27].

Taking into account the active principle of the sensor, pyranometers may be thermoelectric or semiconductor devices which may in turn operate in photoconductive or photovoltaic mode. Within the thermoelectric, two types are distinguished: those that use the junction of two metals with different function work to generate a potential difference proportional to the temperature change (thermopile type) and those that generate a difference of voltage as a result of the change in the material polarization induced by changes in temperature (pyroelectric type).

Pyranometers based on semiconductor devices have as a sensitive element a photodiode, which generates a photocurrent proportional to the solar radiation intensity. This consists of a semiconductor device formed by a P/N junction that generates electron–hole pairs by absorption of photons. These load carriers move by diffusion to the P/N junction, where there is an electric field that moves them towards the electrical contacts, thus generating a photocurrent proportional to the intensity of the incident radiation. The device provides the possibility to operate in photoconductor mode by applying a reverse polarization voltage through the junction, or in photovoltaic mode, without applying polarization voltage.

Due to the electronic noise of the photodiode increases when a polarization voltage is applied, the operation in photovoltaic mode has a better signal-to-noise ratio than in photoconductor mode. The main limitation of photodiode type pyranometers is that their response depends on the spectral content of the radiation, while thermopile type does not [28]. This makes them less effective than the thermopile type for measuring the infrared component of solar radiation. However, they have excellent detectivity and response speed in the visible.

(b) Room temperature sensor

There are various devices used to measure temperature, ranging from different kinds of thermocouples to semiconductor devices in specialized integrated circuits. In this work will be used as temperature sensor a thermistor, which is a device that changes its electrical resistance when changing the temperature. This device was selected because it is very low cost, it has high sensitivity and mechanical stability and easy to operate.

For precise temperature measurements, a resistance vs temperature curve given by the Steinhart–Hart equation is used [29], which is a widely used third-order approximation; this is given by:

$$T = \frac{1}{a + b \ln R + c(\ln R)^3} \tag{3.15}$$

Where a, b and c are called Steinhart–Hart parameters and must be specified for each device. In this expression T is the temperature expressed in Kelvin and R is the resistance expressed in Ohms. The NTC thermistors can be characterized from a parameter known as parameter B, obtained from the reciprocal of expression (3.15), and given by

$$\frac{1}{T} = \frac{1}{T_0} + B \ln\left(\frac{R}{R_0}\right) + c(\ln R)^3 + \ldots \tag{3.16}$$

Where R_0 is the resistance to temperature T_0 of 25 °C, taken as the standard working temperature.

(c) Current transducer: clamp meter

The current clamp is a very useful current measurement instrument that can measure ac and dc current through active conductors without the need to interrupt the circuit.

One of the advantages of this method is that large currents can be measured without the need to disconnect the circuit being measured. Usually an alternating current clamp operates just like a current transformer (CT) by capturing the magnetic flux generated by the current flowing through a conductor. Assuming that the current flowing through the conductor to be measured is the transformer's primary, a current proportional to the primary in the transformer's secondary (coil) is obtained (by electromagnetic induction), which is connected to the instrument measuring circuit. This provides the ac current reading on the screen (in the case of digital current clamps). Figure 3.4 illustrates the basic operation of an AC current clamp.

Current clamps capable to measure dc current use Hall elements for this purpose. Figure 3.5 shows a block diagram of the elements that make up the Hall effect-based clamp. The Hall elements are located in the gap between the two parts of the jaw, such that when a magnetic flux (generated by a dc or ac current) is generated through the jaw, a Hall voltage proportional to the polarization current and the magnetic field generated by the current to be measured.

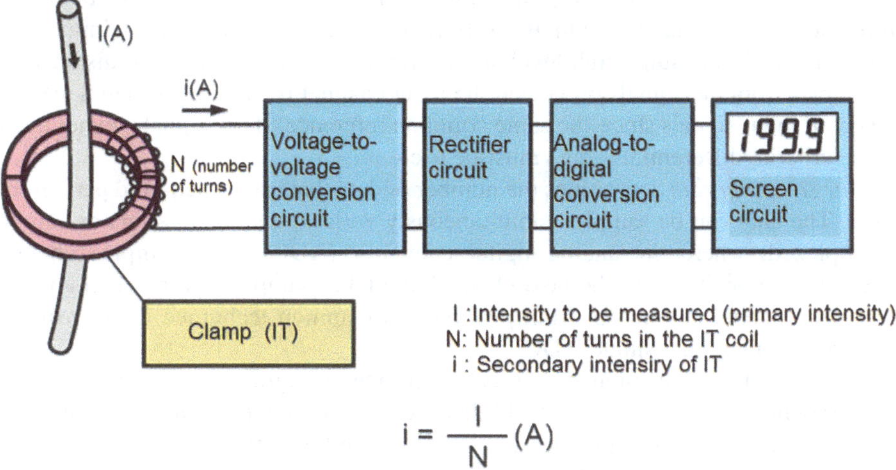

Fig. 3.4 Operating principle of AC clamps

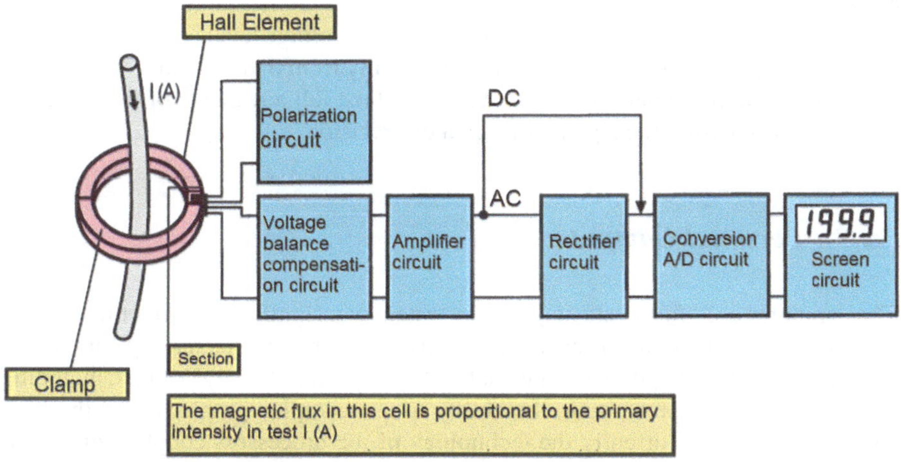

Fig. 3.5 DC current measurement using a Hall effect current clamp

3.2.2 Data Acquisition Hardware

Data acquisition hardware acts as the interface between the computer and the environment. Its main function is to act as a device that digitizes incoming analog signals so that the computer can interpret them.

DAQ device manufacturers provide technical input and output specifications: number of channels, sampling rate, resolution, input range, etc., which are related to the capacity and accuracy of the device.

The number of analog/digital input channels are specified for both the referenced input channels and the differential channels. Referenced inputs are typically used to acquire high level signals (more than 1 V), when the distance of the cables from the signal source and the input channel is small (less than 8 m) and when all input signals share the same common reference. If the signals do not meet these criteria, differential inputs must be used.

The sampling rate determines the number of data that can be acquired per unit of time. The data can be sampled simultaneously with multiple converters or can be multiplexed, where the analog-digital converter (ADC) takes samples on one channel, then switches to the next channel and takes samples again; Then goes to the next channel and so on. Multiplexing is a common technique for measuring several signals with a single ADC.

The resolution is the number of bits that the analog-digital converter (ADC) uses to represent the analog signal. The higher the resolution, the greater the number of divisions in the measuring range, and therefore the lower the change detectable in the voltage. A 3-bit converter, divides the range into 23 divisions. By increasing the resolution to 16 bits, the number of codes of the ADC is increased from 23 to 216 and therefore, an extremely close representation to the original analog signal can be obtained.

The range refers to the maximum and minimum voltage levels that the ADC can acquire. DAQ cards offer selectable ranges (usually from 0 to 10 V or −10 to 10 V), so it can adjust the signal range of that ADC to take full advantage of the available resolution to measure the signal with greater precision.

3.2.3 Personal Computer

The computer used for a data acquisition system can significantly affect the speeds at which data can be acquired continuously. As computer technology improves constantly, the DAQ system can take advantage of computer capabilities, including improved real-time processing, the ability to use complex graphics, and the high data-to-disk speed. Currently the technology of the processors coupled quite well with the performance of the architecture of the different buses and ports. The PCI bus and USB port are standard equipment on most current desktop computers, and have data transfer rates up to 132 Mbytes/s. External and portable computer buses such as PCMCIA and USB offer a flexible alternative to data acquisition systems that use desktop computers, achieving transfer speeds of up to 40 Mbytes/s. When choosing a DAQ device, the data transfer methods supported by the chosen device and the bus transfer rates must be taken into account.

PCs currently have scheduled data transfer capabilities. Transfers to direct memory access (DMA) increase system performance by using dedicated hardware for transferring data directly into system's memory. Using this method the computer is not saturated when moving the data and for that reason is free to perform more complex tasks. The limiting factor for real-time storage of that amount of data is the hard disk. Hard disk access time and fragmentation can significantly reduce the maximum speed at which data is acquired and stored on disk.

3.2.4 *Software*

The software transforms the PC and the DAQ device into a complete data acquisition, analysis and presentation system. Without software to control or handle hardware, the DAQ device will not work or perform properly. Most DAQ applications use a software driver, which is the software layer that programs the DAQ hardware records by managing its operation and integration with computer resources such as the processor and memory. The software driver hides the complicated details of low-level hardware programming, giving the user an easy-to-understand interface.

For the selection of a software driver there are two factors to consider:

1. Available functionality

The functions of DAQ hardware drivers can be grouped into Analog Inputs/Outputs, Digital Inputs/Outputs and timed Inputs/Outputs. Although most drivers expose this basic functionality, it is necessary in addition that the driver has the ability to:

- Test channels without any programming.
- Acquire data without any internal processing.
- Use scheduled input/output data transfers.
- Transfer data to and from the hard disk.
- Perform several functions simultaneously.
- Integrate multiple DAQ devices.
- Easily integrated with various sensors and signal types.

2. The operating system to be used with the driver

The driver must be designed to operate on the different characteristics and capabilities of the operating system. It also requires the ability to easily cross-platform encryption, such as between Windows PC and Macintosh. All of the components of a data acquisition system, software should be the most carefully examined. Because DAQ devices do not have screens, software is the only system's interface. The software is the component that retains all the system's information and therefore that is the element that controls it. The software integrates transducers, signal conditioning, DAQ hardware and analysis within a complete and functional system.

3.3 LabVIEW Software in Virtual Instrumentation

The conventional instruments have a predefined functionality from the factory, where it is designed, produced and assembled. That is, the functionality of this type of instrument is defined by the equipment manufacturer, and not by the user himself. In the "virtual" instrument the PC is used as "instrument" and it is the

Table 3.1 Traditional instruments vs virtual instruments

Traditional instrument	Virtual instrument
Factory set	User defined
Specific functionality with limited connectivity	Unlimited functionality, application-oriented, broad connectivity
Hardware is the main thing	Software is the main thing
High cost/function	Low cost/function, variety of functions
"Closed" architecture	"Opened" architecture
Slow incorporation of new technology	Rapid incorporation of new technologies, thanks to the PC platform
High cost of maintenance	Low cost of maintenance

user himself who through the software defines its functionality and "appearance." That is why it is said that the instrument is "virtualized," since its functionality can be defined again and again by the user and not by the manufacturer.

A virtual instrument consists of a personal computer equipped with powerful programs (software) and hardware such as cards and drivers that together fulfill the functions of traditional instruments. The synergy between them offers advantages that cannot be matched by traditional instrumentation. Therefore, only a PC, a data acquisition card with signal conditioning (PCMCIA, ISA, XT, PCI, etc.) and the appropriate software are required to build a virtual instrument.

In the virtual instrument, software is the main element of the system, unlike the traditional instrument, where the main thing is the hardware. With virtual instruments engineers and scientists are building measurement and automation systems that exactly fit their (user-defined) needs, rather than being limited by traditional (defined by the manufacturer) fixed-function instruments.

Table 3.1 indicates some of the main differences between the traditional instrument and the virtual instrument [30].

Other advantages of virtual instrumentation are that it can also be implemented in laptops, field distributed equipment (RS-485), remote equipment (connected via radio, Internet, etc.) or industrial equipment (NEMA 4X, etc.). There is also a data acquisition card for almost any bus or communication channel in PC (ISA, PCI, USB, serial RS-232/485, parallel EPP, PCMCIA, CompactPCI, PCI, etc.) A driver for almost any operating system (WIN 3.1/95/NT/2000/XP/VISTA, DOS, Unix, MAC OS, etc.).

In conclusion, a virtual instrument can perform the three basic functions of a conventional instrument: acquisition, analysis and presentation of data. However, the virtual instrument additionally allows the following:

- Instrument customization and much more functionality added without incurring in additional costs.
- Establish instrument connectivity with Ethernet.
- Store data in a table or file compatible with MS Excel.
- Add to the instrument a new algorithm or function needed in the experiment.

– Through the accompanying software, the level of adaptability and personalization of the virtual instrument is almost unlimited.

One of the powerful and widely studied tools in the market for implementing instrumentation and industrial control projects is the National Instruments LabVIEW program. This is an entire development environment based on graphical programming with which any desired algorithm can be structured. It has the common elements of any programming language, in addition to the debugging tools that any traditional compiler can offer. It also has a great variety of objects already elaborated that allow to quickly developing any application. In addition, it uses all the terminology, icons, and ideas that are very familiar to engineers, technicians, and scientists of all the areas.

Programs made in LabVIEW are known as virtual instruments (VIs) and are composed of two fundamental parts: the front panel and the block diagram. There exists within the interface of the software a table of utilities or set of pallets that contains all the functions with which modifications to the VI can be made.

The front panel is the graphical interface for the application. This interface collects the user data and displays the output results of the program. It can contain graphical buttons, indicators and controls.

The block diagram contains the graphic source code that constitutes the VI. The VI is programmed in the block diagram to control the functions performed on the inputs that have been created on the front panel. This diagram can include functions and structures starting from the existing libraries. The block diagram and the front panel are strongly linked, i.e., everything that is done with the use of the set of paddles in one can be seen in the other.

LabVIEW VIs follow a data flow model to run each of the programs performed. The block diagram consists of nodes that can be, for example, VIs, structures originated in the paddles or terminals used in the front panel. These nodes are joined by user-drawn connections thanks to the wire tool. The nodes and the wiring define and show the sequence in which the programs develop and the routes of the data flow must follow through each of the stages implemented until arriving at the final result.

One of the great advantages of LabVIEW is its easy integration with signal conditioning hardware, such as SCXI (Signals Conditioned Extensions Interfaces) systems. A SCXI system is a high-performance signal conditioning platform for measurement and automation systems consisting of multi-channel modules installed in a chassis. It is possible to choose from a wide variety of analog input/output modules or digital input/output to precisely match the needs of the application. The NI-DAQ software driver controls every aspect of the DAQ system from configuration to LabVIEW programming, the low-level operating system, and device control.

References

1. R. Poroski, *Proceedings 2nd World Conference on Solar Energy Convertion*, Wien, 1998
2. W.H Bloss, F. Pfisterer, W. Kleinkauf, M. Landau, H. Weber and H. Hullman, *Proceedings 10th European Photovoltaic, Solar Energy Conf.*, Lisbon, 1991
3. D.S. Shugar, *Proceedings 21th IEEE Photov. Specialist Conf.*, 1990 836
4. J.J. Iannucci and D.S. Shugar, *Proceedings 22th IEEE Photov. Specialist Conf.*, 1991
5. A. Jaeger, *A European Roadmap for PV R&D*, E-RMS Spring Meeting, 2003
6. M.A. Green, Recent development in photovoltaics. Sol. Energy **76**, 3–6 (2004)
7. E.J. Wildenbeest, S.W.H. de Haan, N.C. van der Borg, *Recent Test Results of 6 Commercial Inverters for Grid Connected PV Systems*, Proceedings of the First World Conference on Conversion of Photovoltaic Solar Energy, 909, Hawaii, Estados Unidos, 1994
8. H. Haeberlin, F. Kaeser, Ch. Liebl and Ch. Beutler, *Results of Recent Performance and Reliability Tests of the Most Popular Inverters for Grid Connected PV Systems in Switzerland*, Proceedings of the XIII European Congress of Photovoltaic Solar Energy, 585, Nice, France, 1995
9. M. Visiers Guixot, *Inverters for Large Scale PV Plants*. Proceedings of the seminar, Experiences and Perspectives of Large Scale PV Plants, Organized by Task VI of the International Energy Agency–Photovoltaic Power Systems Program, Madrid, Spain December, 1997
10. C.W.A. Baltus, J.A. Eikelboom, R.J.C. van Zolingen, *Analytical Monitoring of Losses in PV Systems*. Proceedings of the XIV European Congress of Solar Photovoltaic Energy, 1547, Barcelona, Spain, 1997
11. P. Schaub, A. Mermoud, O. Guisan, *Evaluation of the Different Losses Involved in Two Photovoltaic Systems*, Proceedings of the XII European Congress of Solar Photovoltaic Energy, 2270, Amsterdam, The Netherlands, 1994
12. W. Coppye, W. Maranda, Y. Nir, L. De Gheselle, J. Nijs, *Detailed Comparison of the Inverter Operation of Two Grid-Connected PV Demonstration Systems in Belgium*, Proceedings of the XIII European Congress of Solar Photovoltaic Energy, 1881, Nice, France, 1995
13. D. Schekulin, G. Schumm, *AC-modules-Technology, Characteristics And Operational Experience*, Proceedings of the XIII European Congress of Solar Photovoltaic Energy, 1889, Nice, France, 1995
14. D. Schekulin, A. Bleil, C. Binder, G. Schumm, *Module-Integratable Inverters in the Power-Range of 100–400 W*, Proceedings of the XIII European Congress of Solar Photovoltaic Energy, 1893, Nice, France, 1995
15. H. Oldenkamp, I.J. de Jong, C.W.A. Baltus, S.A.M. Verhoeven, S. Elstgeest, *Reliability and Accelerated Life Tests of the AC Module Mounted OKE4 Inverter*, Proceedings of the XXV IEEE Photovoltaic Specialist Conference, 1339, Washington, United States, 1996
16. G. Ebest, W. Hiller, U. Knorr, *A New Method and System for the Simulation of Modular Inverters in Solar Façades*, Proceedings of the 25th IEEE Photovoltaic Expert Conference, 1449, Washington, United States, 1996
17. P. Tzanetakis, M. Tsilis, *Design, Construction and Comparative Evaluation of two Portable Photovoltaic Array I-V Curve Tracers Based on Different Operating Principles*, Proceedings of the XI European Congress of Solar Photovoltaic Energy, 1359, Montreaux, Switzerland, 1992
18. Commission of the European Communities, Joint Research Center–Ispra Establishment, *Guidelines for the Assessment of PV Plants. Document C: Initial and Periodic Tests on PV Plants*, Report EUR 16340 EN, 1995
19. A. Rabl, *Active Solar Collectors and their Applications* (Oxford University Press, New York, 1985)
20. *Concerted Actions on PV Systems Technology and Coordination of PV Systems Development*, Task 6 of the JOULE II project with reference 0120, May, 1993
21. H. Wilk, Inverters for Photovoltaic Systems; Chapter 5 of the Book Photovoltaic Systems, edited by the Fraunhofer Institute für Solare Energiesysteme, Freiburg, Germany (1995)

22. M. Jantsch, H. Schmidt, J. Schmid, *Results of the Concerted Action on Power Conditioning and Control*, Proceedings of the XI European Congress of Solar Photovoltaic Energy, 1589, Montreux, Switzerland, 1992
23. M.H.Magnagnan, E. Lorenzo, *On the Optimal Size of Inverters for Grid-Connected PV Systems*. Proceedings of the XI European Congress of Solar Photovoltaic Energy, 1167, Montreux, Switzerland, 1992
24. K. Peippo, P.D. Lund, Optimal sizing of grid-connected PV-systems. Sol. Energy Mater. Sol. Cells **32**, 95 (1994)
25. A. Lester, D.R. Myers, A method for improving global pyranometer measurements by modeling responsivity functions. Sol. Energy **80**(3), 322–331 (2006)
26. D.R. Myers, K.A. Emery, T.L. Stoffel, Uncertainty estimates for global solar irradiance measurements used to evaluate PV performance device. Sol. Cells **27**, 455–464 (1989)
27. L. Arboledas, F.J. Battles, F.J. Olmo, Solar resource assessment by means of silicon- cell. Sol. Energy **5**, 183–191 (1995)
28. Kipp & Zonen. Www.kippzonen.com
29. J.S. Steinhart, S.R. Hart, Calibration curves for thermistors. Deep-Sea Res. Oceanogr. Abstr. **15**(4), 497–503 (1968)
30. retrieved from http://zone.ni.com/devzone/conceptd.nsf/webmain/43DA4F5979907DDD8625 6C1B00510DBA

Chapter 4
Integrated Photovoltaic System Sizing and Economic Evaluation Using RETScreen™ for a Building of 40 Apartments

The growth of PV market had a remarkable rate over the last 10 years even with difficult world economy scenarios and it is getting to be a great source of power generation for the world [1]. After a record increase in 2011, the global PV markets got stabilize in 2012, to grow again in 2013 [2].

Since BIPV offers the possibility to replace part of the traditional building material, with a possible price reduction in comparison to a classic rooftop installation [3, 4], the correct estimation of system level performances, system reliability, and system availability is becoming more important and popular among installers, integrators, investors, and owners; with this purpose several tools and models were developed [1, 5–7]. The combination of different phenomena, such as the solar radiation available on site, the presence of dust, the shadowing or UV radiation over long outdoor exposure, affect in different ways the real performance of BIPV systems and thus the related economic evaluations [8–10].

In 2015, some 60 GWp of new photovoltaic (PV) systems were installed globally [11] bringing the total worldwide installed capacity to nearly 250 GWp, with Asia leading the wave of new installations [12]. Information Handling Services Inc. (HIS) has raised its global solar PV forecasts for 2016 to 65 GWp, and over 70 GWp is expected to be installed in 2019 [13]. By 2020, the cumulative global market for solar PV is expected to triple to around 700 GWp [14].

In addition to being a renewable and pollution free energy generation technology with no moving parts, PV modules can also be integrated into buildings as BIPV systems, adding aesthetic value [15]. When installed in an optimized way, BIPV systems can reduce heat transferred through the envelope and reduce cooling load components decreasing the CO_2 emissions [16]. Apart from some façade installations, the rooftop segment represented more than 23 GWp of total installations in 2015, with projections of more than 35 GWp to be installed by 2018 [12].

There are several studies about photovoltaic energy in Colombia [10, 17, 18] and this one is intended to improve the distributed energy analysis on buildings.

A.J. Aristizábal Cardona et al., *Building-Integrated Photovoltaic Systems (BIPVS)*,
https://doi.org/10.1007/978-3-319-71931-3_4

4.1 Application Analysis

4.1.1 Building Loads Calculation

Effective impedance calculation is performed by the following equation:

$$Z_{ef} = R*\text{Cos}\phi + X*\text{Sin}\phi \tag{4.1}$$

Where:
R: resistance.
X: reactance.
To calculate the voltage regulation the following equation is used:

$$\Delta V\% = \frac{\sqrt{3}*I*l*(R*\text{Cos}\phi + X*\text{Sin}\phi)}{V_0}*100 \tag{4.2}$$

Where I is the electric current, l is the cable length, and Vo the voltage.
In the case of single phase circuits, the following equation is used:

$$\Delta V\% = \frac{2*I*l*(R*\text{Cos}\phi + X*\text{Sin}\phi)}{V_0}*100 \tag{4.3}$$

Equations (4.4, 4.5, 4.6) allow to calculate the active, reactive and apparent power respectively:

$$P = V*I*\text{Cos }\phi \tag{4.4}$$

$$Q = V*I*\text{Sin }\phi \tag{4.5}$$

$$S = V*I \tag{4.6}$$

Meanwhile, the maximum diversified demand for a building can be determined by the following expression:

$$\text{DMD}_{N\text{ Users}} = \frac{\text{Maximum } mathrmdemand \text{ 1 User} \times (N \text{ users})}{f \text{ diversity } N \text{ users}} \tag{4.7}$$

The maximum diversified demand for n years is calculated as:

$$\text{DMD}_{\text{year } n} = \text{DMD}_{\text{year } 0} \times \left(1 + \frac{r}{100}\right)^n \tag{4.8}$$

With maximum diversified demand then transformer capacity is calculated as:

$$\text{S}_{\text{transf}} = \text{S}_{\text{user.res.}} + \text{S}_{\text{common areas}} + \text{S}_{\text{pub.lights}} \tag{4.9}$$

4.1.2 Residential Energy Demand

The cost per kW of conventional energy in Colombia is closely linked to the hydraulic generation and is associated with the level of reservoirs as can be seen in Fig. 4.1, which presents the average costs between 2000 and 2012, and ranging from USD \$0.017 to USD \$0.073 per kWh depending on the greater or lesser amount of water in reservoirs, respectively [19].

As shown in Fig. 4.1, the energy available in the reservoirs varies between 7000 and 15,000 GW from 2000 to 2012. The availability of water reservoirs had been above the energy stock price until 2009 when it was surpassed but then it normalized again in 2010.

Figures 4.2 and 4.3 show the residential energy consumption by social strata in Bogotá between 1998 and 2012.

It can be seen that consumption per strata for 2012 reflects a socioeconomic difference, although the strata 1 and 2 converge. Since 1998 the consumption is falling; but it starts to recover in 2004; especially stratas 1, 2, and 3 [20].

Table 4.1 shows the projected energy demand in Bogota and Cundinamarca until 2020.

The added model predicts an annual growth of energy demand in the region between 3.7 and 4.1% and the sector model between 3.6 and 3.8%. The total annual demand is 14,500 GWh. By 2020 it is predicted that the total energy demand will be between 18,500 and 19,000 GWh. Predictions are close to those of the UPME in the low range for UCP Center with annual growth of 3.1% and 6.7%; to approach a demand of 19,600 GWh in 2020 [20].

Figure 4.4 shows the predictions of power demand for Bogota and Cundinamarca to 2018.

The predictions of power demand apply the energy predictions, the load factor and maximum power from UPME [20]. The increase in power between 2013 and 2018 is 370 MW corresponding to 16%, reaching 2679 MW in 2018.

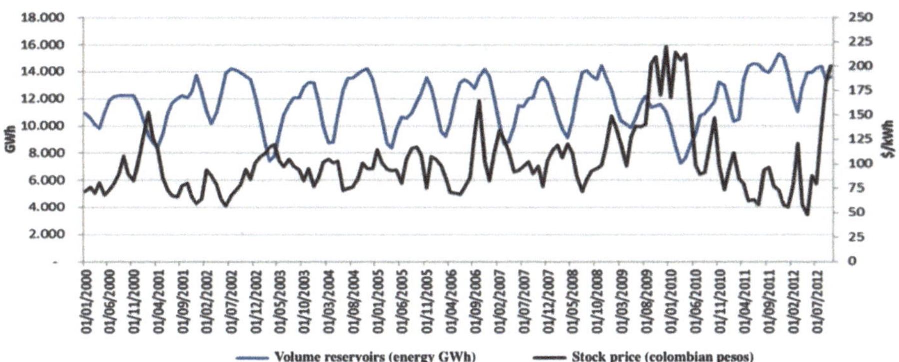

Fig. 4.1 Ratio of stock price with the volume of reservoirs

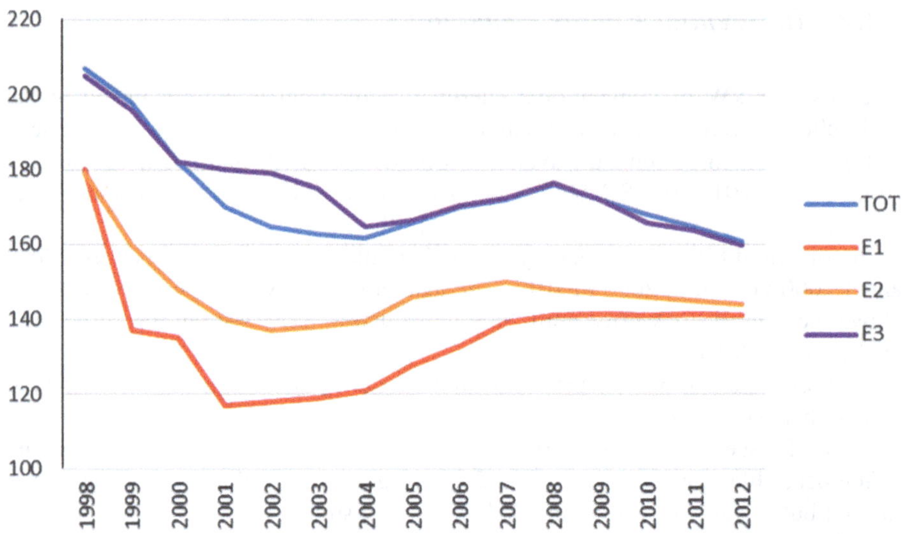

Fig. 4.2 Average monthly consumption (kWh). Residential strata 1, 2 and 3 [20]

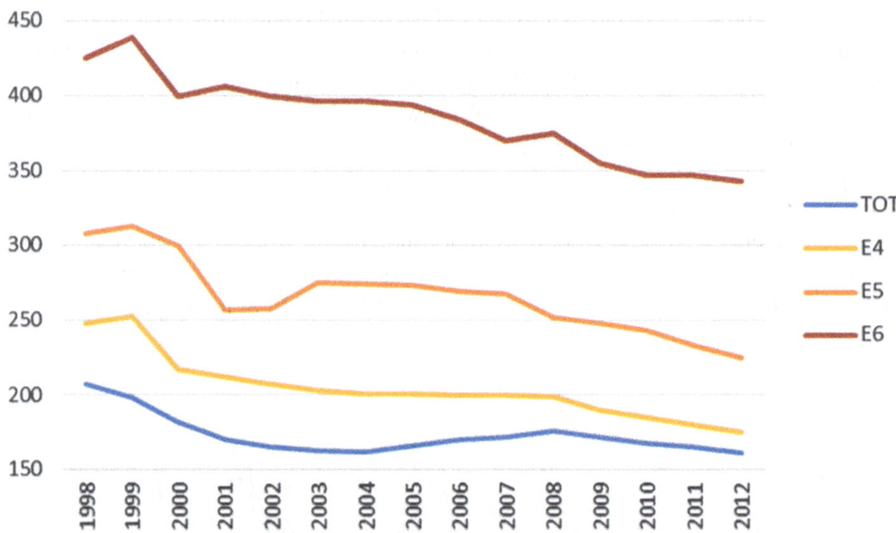

Fig. 4.3 Average monthly consumption (kWh). Residential strata 4, 5 and 6 [20]

Table 4.1 Results of the estimation of energy demand of Bogotá until 2020 [20]

	Model added		Sector model		Upme	
Year	GWh	Incr.%	GWh	Incr.%	GWh	Incr.%
2012	13.940		13.940		13.940	
2013	14.508	4.1	14.363	3.0	14.720	5.6
2014	15.083	4.0	14.895	3.7	15.207	3.3
2015	15.685	4.0	15.453	3.7	15.739	3.5
2016	16.297	3.9	16.031	3.7	16.801	6.7
2017	16.959	4.1	16.645	3.8	17.547	4.4
2018	17.618	3.9	17.267	3.7	18.316	4.4
2019	18.275	3.7	17.896	3.6	19.075	4.1
2020	19.024	4.1	18.583	3.8	19.674	3.1

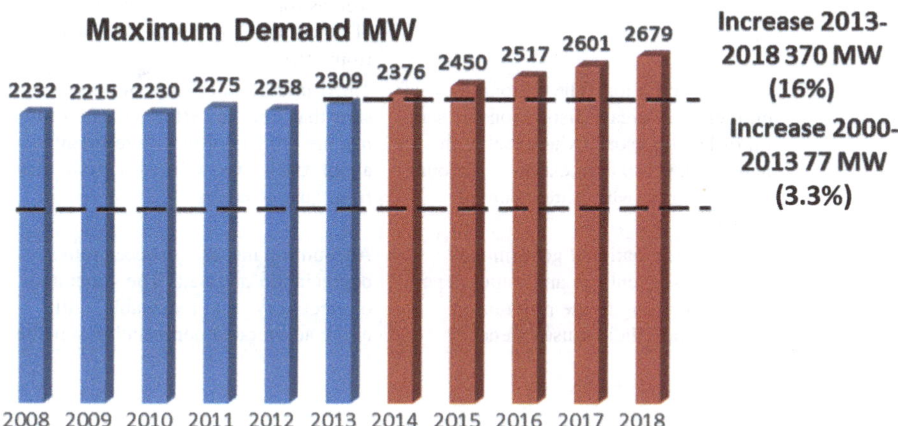

Fig. 4.4 Predictions of power demand for Bogota and Cundinamarca to 2018 [20]

4.1.3 Colombian Law 1715–2014

The Congress of Colombia created the Law 697 of October 3, 2001; where the rational and efficient use of energy, the renewable energy and other provisions are promoted.

In 2014, Law 1715 (May 13th) was promulgated; where the integration of non-conventional renewable energy to the National Energy System (SIN) is regulated.

Table 4.2 presents a comparison of both laws.

Incentives promulgated in Law 1715–2014 represent a significant advance and they will allow promoting new unconventional energy technologies interconnected with SIN. New models of support for such projects are needed, related to the energy purchase in kWh to the company providing the electrical service at prices that even

Table 4.2 Comparison of incentives granted by Law 697–2001 and Law 1715–2014 [21]

Aspect	Law 697–2001	Law 1715–2014
Incentives	For research: The National Government will promote research programs about rational and efficient use of energy through Colciencias	Incentives for non-conventional energy generation: Promote research, development and investment in the field of energy generation with renewables and efficient energy management. Those forced to declare income but directly make investments in this regard, are entitled to deduct from their income annually for the 5 years following the taxable year
	For education: Icetex benefits loans to students who want to pursue careers or specializations about renewable	IVA tax incentive: Equipment, components, machinery and services which are intended for renewable energy, as well as for measurement and evaluation of potential resources, are excluded from IVA
	Public recognition: The national government will create distinctions to natural or legal persons, who excel at the national level in application of rational use of energy; which are awarded annually	Tariff incentive: Natural or legal persons that from the effective date of this law are holders of new investments about renewable, will enjoy exemption from customs duties
	General: The national government establishes incentives and imposes penalties, according to the program of rational and efficient use of energy	Accounting incentive: Accelerated depreciation of assets. The generation of electricity from renewable, will enjoy accelerated depreciation scheme

double the cost of the kWh conventionally generated (hydro or thermal) and the financing of up to 70% of the initial investment as it is currently the case in countries like Germany, Japan and the USA.

For the reasons mentioned above, distributed photovoltaics for residential becomes very important because this type of energy can support residential energy demand in Bogotá and contribute to the reduction of greenhouse gases.

4.2 Sizing and Economic Evaluation

4.2.1 Energy Demand and Photovoltaic Generator Sizing

The analysis is focused on the photovoltaic generation system connected to the distribution network as distributed generation option. The building has 40 apartments with an area of 72 m^2 each and the energy demand is supplied 100% by photovoltaic panels at the common areas of the building. Recommendations of the Colombian Electrical Code NTC 2050 are taken into account.

Table 4.3 Electrical demand of the building and transformer's capacity

40 apartments of 72 m² each		
Load $M^2 = 72\ M^2 \times 32VA/M^2$	2305	NTC 2050 SEC. 220 table 220–3 -b
Load small devices APT 1	3000	NTC 2050 SEC. 220–3 b
Subtotal load	5305	
First 3000 VA × 100%	3000	NTC 2050 SEC. 220 table 220–11
Another ones (5305–3000 = 2305) POR 35%	807	NTC 2050 SEC. 220 table 220–11
Load total	3807	DM
DMD 1 Year =3807×40/3.28	46,423	DMD—1 Year
DMD 8 Years = $46{,}423\times(1 + 3/100)^8$	58,807	DMD—8 Year
Common areas DMD	20,582	Installed capacity (W)
Maximum demand	79,389,45	DMD total
Transformer capacity	112.5 kVA	

Table 4.4 Total energy demand by the common areas of the building

Photovoltaic energy–common areas	
Energy demand-common areas—KWh/month	2460
Power photovoltaic array—KW	25.6

The maximum demand of the building is 79,389,45 kVA as shown in Table 4.3.

The maximum demand for the first 8 years of operation of the building is 58,807 kVA; total electric load of common areas is 20,582 kVA and the transformer required to manage the total demand of the building must be 112.5 kVA. The total current is 313 A and cables selected were 1LXF No. 350 with 3×350 A regulated protections chosen. Table 4.4 shows the adjusted value of the energy demanded by the common areas of the building:

The total demand of the common areas is 2460 kWh/month requiring a 25.6 kW photovoltaic generator.

The cost per kWh for common areas is USD $0.11; this represents a cost of USD $3247.2/year.

4.2.2 Economic Analysis Using RETScreen™

The first step in the analysis using RETScreen, is to enter the geographical location of the city in which the PV array will be installed to allow the software determine the potential of solar energy in the region.

Bogota is located at 4.7°N latitude, −74.1°E longitude and 2546 m above sea level.

Figure 4.5 shows the weather conditions of Bogota according to RETScreen.

The monthly environmental data indicate that Bogota has an average of 13.3 °C/month; a relative humidity of 81.3%/month and a daily solar radiation of 4.27 kWh/m²

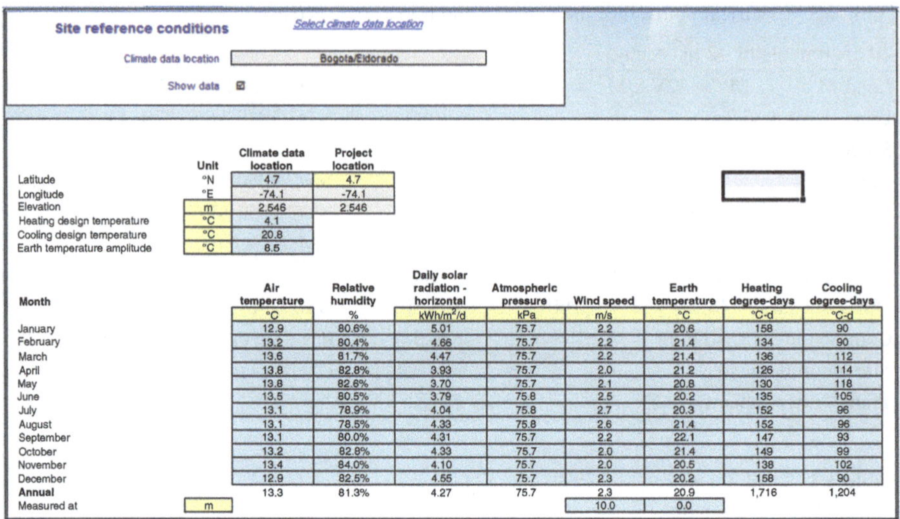

Site reference conditions

Select climate data location

Climate data location: Bogota/Eldorado

Show data ☒

	Unit	Climate data location	Project location
Latitude	°N	4.7	4.7
Longitude	°E	-74.1	-74.1
Elevation	m	2.546	2.546
Heating design temperature	°C	4.1	
Cooling design temperature	°C	20.8	
Earth temperature amplitude	°C	8.5	

Month		Air temperature	Relative humidity	Daily solar radiation - horizontal	Atmospheric pressure	Wind speed	Earth temperature	Heating degree-days	Cooling degree-days
		°C	%	kWh/m²/d	kPa	m/s	°C	°C-d	°C-d
January		12.9	80.6%	5.01	75.7	2.2	20.6	158	90
February		13.2	80.4%	4.66	75.7	2.2	21.4	134	90
March		13.6	81.7%	4.47	75.7	2.2	21.4	136	112
April		13.8	82.8%	3.93	75.7	2.0	21.2	126	114
May		13.8	82.6%	3.70	75.7	2.1	20.8	130	118
June		13.5	80.5%	3.79	75.8	2.5	20.2	135	105
July		13.1	78.9%	4.04	75.8	2.7	20.3	152	96
August		13.1	78.5%	4.33	75.8	2.6	21.4	152	96
September		13.1	80.0%	4.31	75.7	2.2	22.1	147	93
October		13.2	82.8%	4.33	75.7	2.0	21.4	149	99
November		13.4	84.0%	4.10	75.7	2.0	20.5	138	102
December		12.9	82.5%	4.55	75.7	2.3	20.2	158	90
Annual		13.3	81.3%	4.27	75.7	2.3	20.9	1,716	1,204
Measured at	m					10.0	0.0		

Fig. 4.5 Environmental data from Bogota

The energy model used in RETScreen has an electricity export rate that corresponds to the projection of energy sales by the photovoltaic generation system $/MWh.

Based on the price of residential energy sector in Bogota, export rate for the electricity generation system would be USD $0.185/KWh or USD $185 MWh (USD $1 = 2954 COP, October-2017); with the ability to export 31.2 MWh of power to the grid a year.

The technical data of the PV generator is selected through the library that has the RETScreen software. 86 solar panels of 300 Wp each, were selected, Canadian Solar brand, polycrystalline type, for a total power generation capacity of 25.8 kW, with an efficiency of 15.6% per panel. The two inverters selected have an operating capacity of 10 KW each and an efficiency of 98%. Both devices have 10% losses *see* Fig. 4.6.

The costs associated with the application stages are: system design, purchasing and procurement of equipment and materials, procurement and implementation and O&M system.

The overall cost of the application implemented with 25.8 kW was USD $38,203, with the following characteristics:

- Initial costs:

In the initial costs are included: Feasibility, development, engineering, power system and incidentals. This value is USD $38,203 corresponding to 100% of the application.

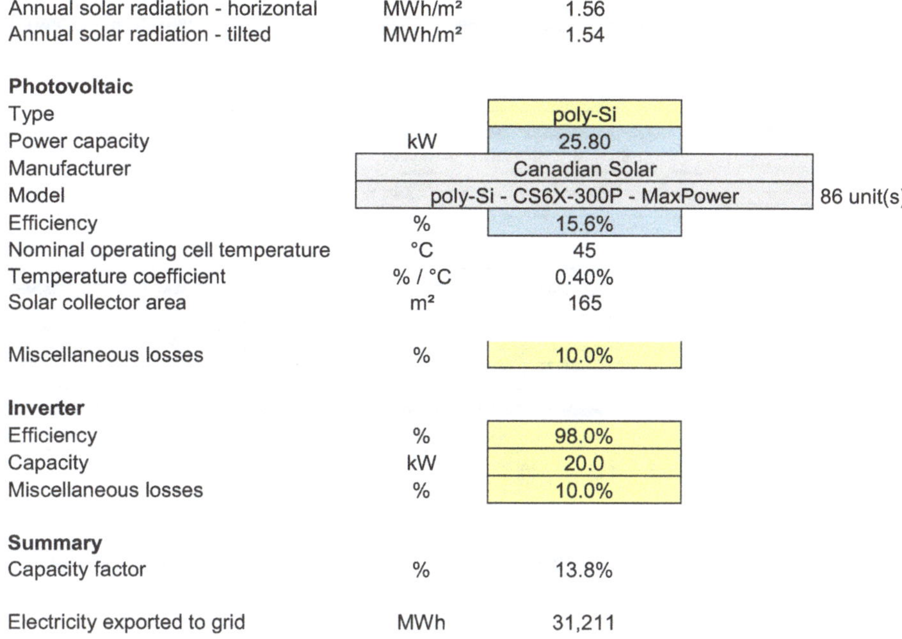

Fig. 4.6 Technical data of solar panels and selected inverters

- Annual costs:

They correspond to that associated with the operation and maintenance O&M cost of the system, whose annual projection was USD $258.

- Period cost:

It refers to the projected costs in years, where the life or replacement of parts involved in the application is determined. In this case the life of inverters is 13 years. The projection of costs for the inverters in the year 13th was USD $6609.

The analysis of greenhouse gases emissions used a base case with diesel generation. This case generates an emission factor of 0.933 tCO_2/MWh and a contribution of 29.1 tCO_2/year from fuel consumption to generate 31 MWh.

Thanks to the use of photovoltaic panels, the emission factor was 0.00 tCO_2/MWh and the contribution was 0.00 tCO_2 for the generation of 31 MWh.

Figure 4.7 shows the CO_2 emissions reduction by using photovoltaic solar energy.

The reduction of total annual emissions of greenhouse gases is 29.1 tCO2, equivalent to 12.503 l of gasoline.

Finally, Fig. 4.8 shows the cumulative cash flows.

The initial economic investment of the application is USD $38,203 and it is recovered in a time period of 6.9 years and energy savings for generation and/or sale of annual energy for 31 MWh at a cost of USD $ 0.185/KWh equivalent to

RETScreen Emission Reduction Analysis - Power project

Emission Analysis

	Global warming potential of GHG	
○ Method 1		
◉ Method 2	25 tonnes CO2 = 1 tonne CH4	(IPCC 2007)
○ Method 3	298 tonnes CO2 = 1 tonne N2O	(IPCC 2007)

Base case electricity system (Baseline)

Fuel type	Fuel mix %	CO2 emission factor kg/GJ	CH4 emission factor kg/GJ	N2O emission factor kg/GJ	Electricity generation efficiency %	T&D losses %	GHG emission factor tCO2/MWh
Diesel (#2 oil)	100.0%	69.3	0.0019	0.0019	30.0%	10.0%	0.933
Electricity mix	100.0%	256.8	0.0070	0.0070		10.0%	0.933

☐ Baseline changes during project life

Base case system GHG summary (Baseline)

Fuel type	Fuel mix %	CO2 emission factor kg/GJ	CH4 emission factor kg/GJ	N2O emission factor kg/GJ	Fuel consumption MWh	GHG emission factor tCO2/MWh	GHG emission tCO2
Electricity	100.0%	256.8	0.0070	0.0070	31	0.933	29.1
Total	100.0%	256.8	0.0070	0.0070	31	0.933	29.1

Proposed case system GHG summary (Power project)

Fuel type	Fuel mix %	CO2 emission factor kg/GJ	CH4 emission factor kg/GJ	N2O emission factor kg/GJ	Fuel consumption MWh	GHG emission factor tCO2/MWh	GHG emission tCO2
Solar	100.0%	0.0	0.0000	0.0000	31	0.000	0.0
Total	100.0%	0.0	0.0000	0.0000	31	0.000	0.0
				T&D losses			
Electricity exported to grid	MWh	31			0	0.933	0.0
						Total	0.0

GHG emission reduction summary

	Base case GHG emission tCO2	Proposed case GHG emission tCO2	Gross annual GHG emission reduction tCO2	GHG credits transaction fee %	Net annual GHG emission reduction tCO2
Power project	29.1	0.0	29.1		29.1

Net annual GHG emission reduction 29.1 tCO2 is equivalent to 12.503 Litres of gasoline not consumed

Fig. 4.7 Analysis of CO_2 emissions avoided by the use of solar energy

USD is $ 5774/year. After 6.9 years the generation or sale of power is reflected in net income, with a small curve in Fig. 4.8 due to the replacement of inverters. The net present value (NPV) for the application to 25 years is USD $ 74,982. It is important to clarify that economic analysis includes the benefits granted by Colombian Law 1715 by eliminating the payment of taxes on solar panels and inverters of the application. This reflects that investing in a generation system based in photovoltaics, is profitable with positive profit margins after 6.9 years of operation.

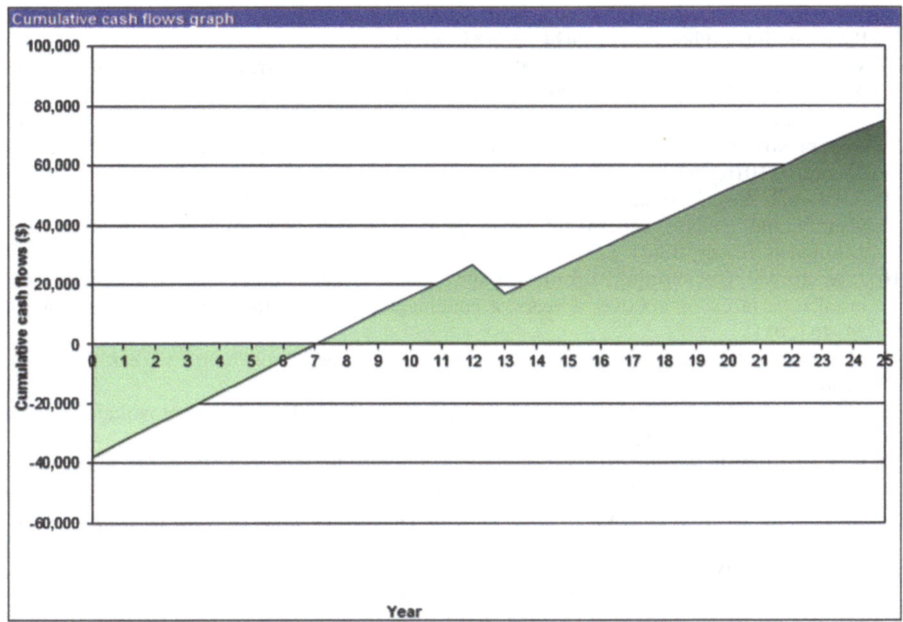

Fig. 4.8 Analysis of cumulative cash flows

In this application the feasibility of installing a photovoltaic system was studied as distributed generation at technical and economic levels. As for the technical feasibility, previously it showed that this depends directly on the energy demand of the building. Due to the high energy demand of the building, a photovoltaic solar energy to supply 100% of the demand of the common areas was used: 86 solar panels of 300 W each (25.8 kW) are required to meet demand.

References

1. N. Aste, C. Del Pero, F. Leonforte, The first Italian BIPV project: Case study and long-term permormance analysis. Sol. Energy **134**, 340–352 (2016)
2. EPIA, (2015). Global market outlook for solar power 2015–2019 [WWW Document]. 2015 <http://resources.solarbusinesshub.com/images/reports/104.pdf>
3. A. Campoccia, L. Dusonchet, E. Telaretti, G. Zizzo, An analysis of feed'in tariffs for solar PV in six representative countries of the European Union. Sol. Energy **107**, 530–542 (2014.) https://doi.org/10.1016/j.solener.2014.05.047
4. T. James, A. Goodrich, M. Woodhouse, R. Margolis, S. Ong, Building- integrated photovoltaics (BIPV) in the residential sector: An analysis of installed rooftop system prices. Energy **50**, 1–39 (2011)
5. V.J. Chin, Z. Salam, K. Ishaque, Cell modelling and model parameters estimation techniques for photovoltaic simulator application: A review. Appl. Energy **154**, 500–519 (2015.) https://doi.org/10.1016/j.apenergy.2015.05.035

6. V. Lo Brano, G. Ciulla, M.D. Falco, Artificial neural networks to predict the power output of a PV panel. Int. J. Photoenergy **2014**, 193083 (2014)
7. W. Zhou, H. Yang, Z. Fang, A novel model for photovoltaic array performance prediction. Appl. Energy **84**, 1187–1198 (2007). https://doi.org/10.1016/j.apenergy.2007.04.006
8. V. Sharma, S.S. Chandel, Performance and degradation analysis for long term reliability of solar photovoltaic systems: A review. Renew. Sust. Energ. Rev. **27**, 753–767 (2013.) https://doi.org/10.1016/j.rser.2013.07.046
9. A. Chica, F. Rey, J. Aristizábal, Application of autoregressive model with exogenous inputs to identify and analyse patterns of solar global radiation and ambient temperature. Int. J. Ambient. Energy **33**(4), 177–183 (2012)
10. J. Aristizábal, G. Gordillo, Performance and economic evaluation of the first grid–connected installation in Colombia over 4 years of continuous operation. Int. J. Sustain. Energ. **30**(1), 34–46 (2011)
11. M. Osborne, *IHS Remains Cautions on PV Market Demand Growth, in: PV-Tech* (PV-Tech, London, 2015)
12. EPIA, *Global Market Outlook for Photovoltaics 2014–2018* (European Photovoltaic Industry Association, Brussels, Belgium, 2014), p. 60
13. B. Beetz, IHS Increases 2015 PV forecast to 59 GW, 2016 to 65 GW, in PV Magazine—Photovoltaic Markets & Technology (2015). Accessed in February, 2016. Available at http://www.pv-magazine.com/news/details/beitrag/ihs-increases-2015-pv-forecast-to-59-gw–2016-to-65-gw 100021513/#axzz45C8rLdVq
14. GTM, Global PV Demand Outlook 2015–2020: Exploring Risk in Downstream Solar Markets (2015). Accessed in January, 2016. Available at. https://www.greentechmedia.com/research/report/global-pv-demand-outlook-2015–2020
15. D. Prasad, M. Snow, *Designing with Solar Power-a Source Book for Building Integration Photovoltaics (BiPV)* (Images Publishing, Australia, 2002)
16. M.S. ElSayed, Optimizing thermal performance of building-integrated photovoltaics for upgrading informal urbanization. Energ. Buildings **116**, 232–248 (2016)
17. A.J. Aristizábal, D.C. Sierra, J.A. Hernandez, Lyfe cycle assessment applied to photovoltaic energy a review. J. Electrical Electron. Eng. **11**(5), 6–13 (2016)
18. J. Aristizábal, Virtual instrumentation applied to identifying parameters of solar radiation and ambient temperature using autoregressive modeling with exogenous inputs. J. Electrical Electron. Eng. **11**(4), 53–58 (2016)
19. Unidad de Planeamiento Minero Energético and Banco Interamericano de Desarrollo, Integración de las energías renovables no convencionales en Colombia (2015)
20. A. Martínez, E. Afanador, J. Núñez, R. Ramírez, J. Zapata, T. Yépez, *Análisis de la Situación Energética de Bogotá y Cundinamarca* (Centro de Investigación Económica y Social – FEDESARROLLO, Bogotá, 2013)
21. Q.S. Grace, *Cuadro Comparativo Entre la Ley 697 de 2001 y la Ley 1715 de 2014* (Universidad de Bogotá Jorge Tadeo Lozano, Bogotá, 2014)

Chapter 5
BIPVS Sizing and Implementation at the Universidad de Bogotá Jorge Tadeo Lozano

The installations connected to the grid must satisfy the energy needs of the user and be compatible with the parameters of the grid. These facilities seek an energy production to obtain economic results in the initial investment return.

Basically the production of a BIPVS is defined by the technical characteristics of the place, the photovoltaic generator, the energy conversion equipment or the inverter and the load to be demanded.

BIPVS presents capture losses caused by shadows and other environmental factors, also present system losses caused by technical and electrical factors (equipment, conductors, transformers, etc.). All these various factors affect the energy efficiency of the photovoltaic system.

For BIPVS sizing, it is essential to know the electrical characteristics of the PV modules supplied by the manufacturer. The data sheet of the PV modules show the following electrical information: Open circuit voltage VOC; Short Circuit Current, ISC; Maximum power, Pm, and TNOC (nominal operating temperature of the cell). These parameters are determined by the manufacturer for each individual module from measurements of the I–V and P–V curves performed under standard conditions defined by the IEC 1215 standard.

5.1 BIPVS Sizing

The sizing of any photovoltaic system (autonomous or interconnected) is done by energy; the two main components of the BIPVS, the generator and inverter, are designed to determine how much energy (in kWh) must be produced by the system in a given time, usually 1 year.

The calculation of the peak power to be supplied by the PV generator to supply a given demand is calculated by means of the following relation

A.J. Aristizábal Cardona et al., *Building-Integrated Photovoltaic Systems (BIPVS)*,
https://doi.org/10.1007/978-3-319-71931-3_5

$$\text{PGPV} = \frac{\sum_{i=1}^{12} \frac{E_i}{\text{HSS}_i N_i \text{PR}}}{12} \quad (5.1)$$

Where E_i is the average monthly solar electricity production (in kWh/month) consumed in the facility, and is generally the expected energy value in the year, divided by twelve; HSS is the number of hours of average monthly standard solar irradiance of the locality; N_i is the number of days in the respective month, and PR is the system performance factor.

In order to achieve greater reliability in the operation of a PV system, a performance factor PR (Performance Power) must be entered. This performance factor is used in such a way that the system is oversized by a value around 10–20%. Then, PR generally has values between 0.7 and 0.9. This factor compensates for irregularities in generation capacity due to effects of temperature, dust, connections, etc.

The other equipment or component of great interest for the development and sizing of PVS is the inverter which converts DC to AC. The suppliers of this energy conversion equipment must ensure compliance with the international standards that guarantee the proper BIPVS functioning [1].

Other important features of the inverters are that they have an electronic MPPT control system (maximum power point follower) to extract the maximum power of the PV panels, varying their maximum operating point; Low influence of harmonic distortion, compatibility with different PV module technologies, operating capacity with other inverters in parallel, secure protection system when connected to external network, among others.

For the inverter sizing it is necessary to define its sizing factor (FSI) as shown in the reviewed relation (14):

$$\text{FSI} = \frac{P_{\text{max, Inverter}}}{P_{\text{nominal, Generator}}} = \frac{P_{\text{max, I}}}{P_{\text{nom, G}}} \quad (5.2)$$

In order to carry out the sizing of this equipment, the generator-inverter relative size is defined so that the photovoltaic generator is oversized between 0.8 and 1 with respect to the inverter according to [2–4], due to the following technical reasons:

- Reduced system cost (smaller inverter) without significantly affecting energy efficiency.
- The rated power of the generator corresponds to standard conditions, which are not common in practice.
- The nominal power of the inverter is the output power; its losses must be taken into account.
- The daily variation of the irradiance throughout the day makes the system most of the time to operate in conditions of low power with respect to the nominal.

Finally, the sizing that was developed and adjusted according to the characteristics required and to the budget available for the BIPVS, is presented below:

The PV generator is composed of modules connected in series and in parallel, with the following electrical characteristics of open circuit voltage Voc, Isc short circuit current, and maximum power P_{max} specific of each module.

Depending on the configuration, the voltage and current corresponding to the maximum power of the generator are calculated. In this case it has a total power as initial condition of 6000 W.

Table 5.1 shows the electrical characteristics of the selected module.

Thus, in order to meet the 6000 W condition and because each PV panel is 250 W, a total of 24 panels are required. With respect to the inverter and using eq. (5.2) with an FSI of 0.85, it gets a maximum power of the inverter of 5100 W. A SunnyBoy 5000 W inverter is selected thanks to its reliability. The following table shows the inverter's characteristics:

As shown in Table 5.2, the DC voltage required to activate the MPPT circuit of the inverter must be in the range of 175–480 V. According to Table 5.3, each panel has a voltage of 38 V. In this way, the 24 solar panels will be connected as follows: two parallel branches of 12 panels connected in series; as shown in Fig. 5.1.

Finally the sizing of the photovoltaic generator is defined according to the following electrical parameters as shown in Table 5.3.

Table 5.1 Electrical characteristics of the selected module

PV Module	Trina solar TSM-PA05.08
Electrical data at STC	
Maximum power (P_{max}) [W]	250
Voltage at maximum power (V_{mpp}) [V]	30.3
Current at maximum power (I_{mpp}) [A]	8.27
Open circuit voltage (V_{oc}) [V]	38
Short circuit current (I_{sc}) [A]	8.79
Module efficiency (nm) [%]	15.3
Electrical data at NOCT	
Maximum power (P_{max}) [W]	186
Voltage at maximum power (V_{mpp}) [V]	28
Current at maximum power (I_{mpp}) [A]	6.65
Open circuit voltage (V_{oc}) [V]	35.2
Short circuit current (I_{sc}) [A]	7.1
Temperature coefficients	
Nominal operating temperature of the cell [°C]	44
Maximum power P_{max} [%/°C]	−0.41
Open circuit voltage V_{oc} [%/°C]	−0.32
Short circuit current I_{sc} [%/°C]	0.05

Table 5.2 Technical characteristics of the selected inverter

PV Inverter	Electrical data	
Brand	Sunny boy	
Model	SB 5000TL-US-22	
Serial/number (S/N)	1913046247	
Output maximum power (W)	5000	
Operating voltage (VAC)		
Minimum	183	211
Nominal	208	240
Maximum	229	264
Operating frequency (HZ)		
Minimum	59.3	
Nominal	60	
Maximum	60.5	
Output maximum current (A)	22	
Power factor (PF)	1	
Operating voltage (VDC)		
Minimum	175	
Maximum	480	
Operating temperature (°c)		
Minimum	40	
Maximum	60	
Degree of protection (IP)	54	
Connection	2F + T	

Table 5.3 BIPVS electrical characteristics at the UJTL

BIPVS	Electrical data
Coordinates—Bogotá DC—Colombia	4° N, 74° W
Trina solar modules TSM-PA05.08	
Generator maximum power [Wp]	6000
Number of series modules	12
Number of parallel modules	2
Generator open circuit voltage [V]	456
Generator maximum power voltage	363.6
Generator shot circuit current [A]	17.58
Total number of modules	24
Performance power (PP)	0.7–0.9
Dimensioning factor (DIF)	0.83
Inverter power—SunnyBoy-SMA [W]	5000
AC inverter configuration	Biphasic—2F + T
DC inverter configuration	L+ L− T
Protection—rail type breaker [A]	2×30
Metallic structure	Electrostatic painting
Modules tilt angle	15°
Orientation	South

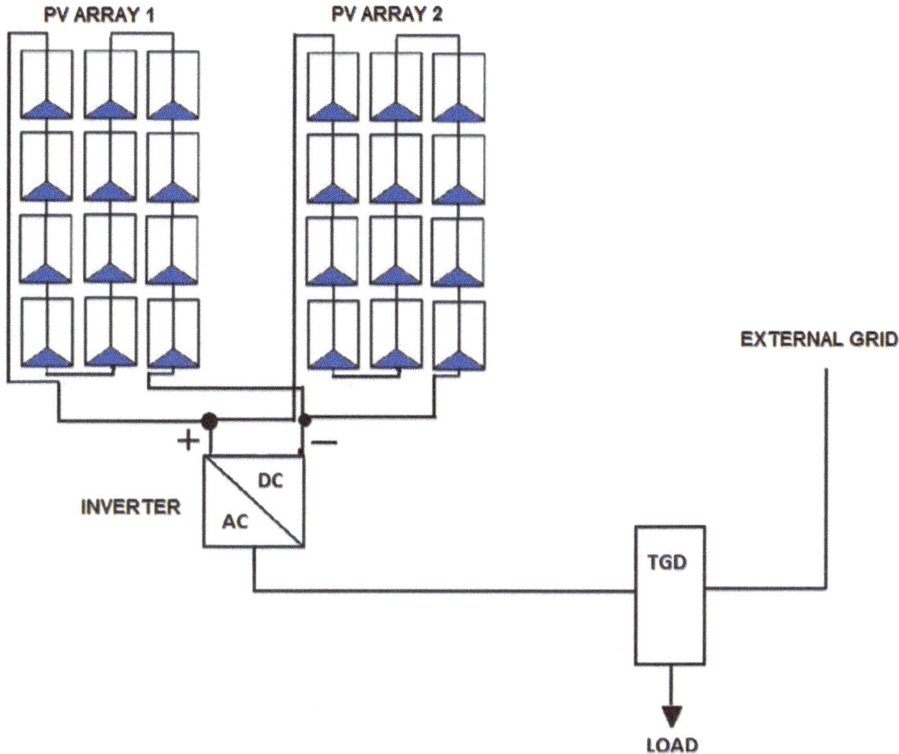

Fig. 5.1 FV arrangement Installed at the UJTL

References

1 IEEE Std 929-2000, Recommended Practice for Utility Interface of Photovoltaic (PV) Systems

2 M. Jantsch, H. Schmidt, J. Schmid, *Results of the Concerted Action on Power Conditioning and Control*, Proceedings of the XI European Congress of Solar Photovoltaic Energy, 1589, Montreux, Switzerland (1992)

3 M.H. Magnagnan, E. Lorenzo, *On the Optimal Size of Inverters for Grid-Connected PV Systems*, Proceedings of the XI European Congress of Solar Photovoltaic Energy, 1167, Montreux, Switzerland (1992)

4 K. Peippo, P.D. Lund, Optimal sizing of grid-connected PV-systems. Sol. Energy Mater. Sol. Cells **32**, 95 (1994)

Chapter 6
Method for Calculating Quantum Efficiency and Spectral Response of Solar Cells Using LabVIEW

This chapter presents the classical procedures for calculating the short-circuit spectral current density for the base and for the emitter of a solar cell. Subsequently, the quantum efficiency and the spectral response are calculated. The method is performed with virtual instrumentation through LabVIEW's graphical programming language.

The developed program processes the wavelength data, the absorption coefficient, the radiation spectrum and the reflection coefficient to develop the analysis mentioned and allows to export the obtained information to files in. xls format.

It is found for a single silicon's solar cell; an emitter short-circuit spectral current density of 24.3 mA/cm^2 (maximum value) and 61.6 mA/cm^2 (maximum value) for the base. The internal quantum efficiency reached 97% for the 0.65 μm; while the internal spectral response recorded 610 mA/W at 1.45 μm.

Photovoltaic technology is a promising renewable energy source that converts sunlight into electricity, with great potential to contribute significantly to solve the future energy problem facing mankind.

To date, solar cells made from semiconductor materials dominate the commercial market, those of crystalline silicone in an 80% participation; the remaining 20% is mostly thin film technology, such as CdTe and CIS and CIGS [1]. The first are semiconductors of prohibited energy band or indirect gap that normally require an absorption layer of 300 mm of thickness and the costs of materials and processing are very high. The latter contain elements that are toxic with low abundance in nature. However, the $CuIn_{12x}Ga_xSe_2$ compound makes it possible to manufacture thin film solar devices with better performance, since it exhibits an efficiency of 20%, but is 1.4 times more expensive than the CdTe and amorphous Si devices [2]. An ecological and low-cost alternative for these solid state devices is the dye-sensitized solar cell (DSC) [3, 4] and the solar cells of polymeric materials that have reached efficiencies close to 9% (Solar Cell Efficiency Tables, Martin Green), whose preparation costs are lower than those of other dominant technologies in the market and its structure allows automated manufacturing.

A.J. Aristizábal Cardona et al., *Building-Integrated Photovoltaic Systems (BIPVS)*,
https://doi.org/10.1007/978-3-319-71931-3_6

The crystalline silicon solar cells manufactured by solar grade silicon material from a metallurgical process, have a lower performance than those made by polycrystalline silicon from a chemical process; but the former have a higher open-circuit voltage than the latter, in low-light conditions [5].

Solar grade silicon material from a metallurgical process represents a cost-effective way for the production of crystalline silicon solar cells; but it is a complex system of semiconductors: its quality suffers from the presence of impurities of transition metals, oxygen, carbon and other complexities [6].

The characterization processes of solar cells represent a valuable source of information to improve the optical and electrical processes of the materials in order to increase the performance of the photovoltaic devices, and in this way; achieve a reduction in costs at the industrial production level.

Virtual instrumentation systems are extensively used in industry and teaching laboratories for experimentation, testing, control, and performance analysis of data acquired through measurements and for simulation and validation processes [7–9].

The process presented in this chapter is carried out using the LabVIEW 2014 software package applying classic definitions of solar cells on a previously characterized silicon solar cell; but can also be used for any solar cell regardless of the type of technology, for which the information required by the developed application is available.

6.1 Initial Considerations

6.1.1 Analytical Model of the Solar Cell

The solar cell is a device manufactured to convert sunlight directly into electricity, through the physical phenomenon called photovoltaic effect [10]. Solar cells do not need to be recharged like a battery; some have been in continuous operation on the ground or in space for more than 30 years [11].

The solar cell has the ability to absorb light and deliver some of the energy from the absorbed photons to the electric current carriers (electrons and holes). A semiconductor separates and collects the charge carriers and conducts the generated electrical current preferably in a specific direction. Thus, a solar cell is simply a semiconductor diode that has been designed and constructed to efficiently absorb and convert the energy of sunlight into electrical energy [12]. A conventional solar cell is shown in Fig. 6.1. Sunlight hits the surface of the solar cell. A wire mesh makes electrical contact with the cell and allows the light to reach the semiconductor through the mesh lines and be absorbed and converted into electrical energy. An antireflective layer between the mesh lines increases the amount of light transmitted to the semiconductor. A diode is formed when a semiconductor type n and a type p are grouped together to form a joint. This is typically achieved through processes of diffusion or implantation of specific impurities (dopants) or by

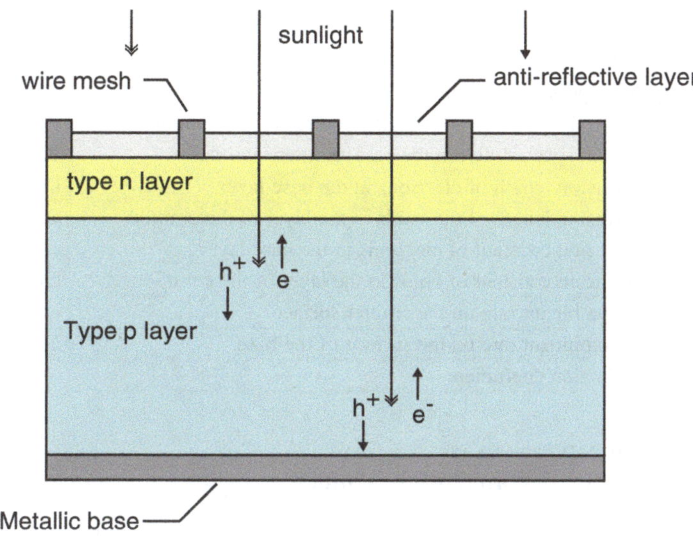

Fig. 6.1 Schematic representation of a conventional solar cell. It represents the creation of electron–hole pairs, e^- and h^+, respectively [11]

deposition processes. The other electrical contact of the diode is formed by a metallic layer at the back of the solar cell [11].

The phenomena for electricity generation in the solar cell can be summarized as: absorption of solar radiation (absorption of photons by electrons), separation of charge carriers at the juncture, collection of these carriers in the device contacts and transport to the different types of power conditioning and energy storage according to the determined final use. For terrestrial applications, each stage of this sequence critically depends on the others mentioned. For example, photovoltaic electricity generation must be coupled with an effective storage process in order to compensate for the intermittent nature of solar radiation [13].

The calculation of the photogenerated current of a solar cell at a given wavelength requires the solution of a set of four differential equations, including continuity and current equations for both minority and majority carriers and the Poisson equation [14].

6.1.2 *Spectral Short-Circuit Current Density*

It corresponds to the current per unit area and per unit wavelength incident on the solar cell.

The model described in [14] delivers the value of the photocurrent collected by a solar cell surface of 1 cm^2 when exposed to monochromatic light.

Table 6.1 Principal parameters involved in the analytical model [14]

Symbol	Name	Units
A	Absorption coefficient	cm^{-1}
Φ_0	Spectral flux of photons on the surface of the emitter	Photon/cm^2 μms
Φ'_0	Photon spectral flux in the base–emitter interface	Photon/cm^2 μms
L_n	Diffusion length of electrons in the base layer	Cm
L_p	Diffusion length of the voids in the layer of the emitter	Cm
D_n	Diffusion constant of electrons in the base layer	cm^2/s
D_p	Diffusion constant of voids in the layer of the emitter	cm^2/s
S_e	Recombinant rate on the emitter surface	cm/s
S_p	Recombinant rate on the surface of the base	cm/s
R	Reflection coefficient	–

The analytical expressions are:
Spectral short-circuit current of the emitter:

$$J_{scE}(\lambda) = \frac{q\alpha\phi_0(1-R)L_p}{\left(\alpha L_p\right)^2}\left[-\alpha L_p e^{-\alpha W_e} + \frac{S_e\frac{L_p}{D_p} + \alpha L_p - e^{-\alpha W_e}\left(S_e\frac{L_p}{D_p}\,\mathrm{Ch}\frac{W_e}{L_p} + \mathrm{Sh}\frac{W_e}{L_p}\right)}{\mathrm{Ch}\frac{W_e}{L_p} + S_e\frac{L_p}{D_p}\,\mathrm{Sh}\frac{W_e}{L_p}}\right] \tag{6.1}$$

Spectral short-circuit current density of the base:

$$J_{scB}(\lambda) = \frac{q\alpha\phi'_0(1-R)L_n}{(\alpha L_n)^2 - 1}\left[-\alpha L_n - \frac{S_b\frac{L_n}{D_n}\left(\mathrm{Ch}\frac{W_b}{L_n} - e^{-\alpha W_b}\right) + S_h\frac{W_b}{L_n} + \alpha L_n e^{-\alpha W_b})}{\mathrm{Ch}\frac{W_b}{L_n} + S_b\frac{L_n}{D_n}\,\mathrm{Sh}\frac{W_b}{L_n}}\right] \tag{6.2}$$

where the main parameters of Eqs. (6.1) and (6.2) are defined in Table 6.1:

6.1.3 Photon Spectral Flow

The photons spectral fluxφ_0 received on the front surface of the emitter of a solar cell is easily related to the spectral radiation and the wavelength, taking into account that the spectral radiation is the power per unit area and wavelength unit.

As shown in Eq. (6.3):

$$\phi_0 = 10^{16}\frac{I_\lambda \lambda}{19.8} \tag{6.3}$$

With $I\lambda$ (I—solar radiation and λ—wavelength) in W/m^2μm and λ in μm.
For the region of the base, we have the Eq. (6.4):

$$\phi' = \phi_0 e^{-\alpha W_e} = e^{-\alpha W_e} 10^6 \frac{I_\lambda \lambda}{19.8} \tag{6.4}$$

In this way, the total short-circuit spectral current density (Eq. 6.5) is the sum of Eqs. (6.1) and (6.2):

$$J_{SC\lambda} = J_{scE(\lambda)} + J_{scB(\lambda)} \tag{6.5}$$

6.1.4 Total Short Circuit Current

It corresponds to the total contribution of current to all wavelengths of incident light in the solar cell.

This is described by Eq. (6.6):

$$J_{SC} = \int_0^\infty J_{sc\lambda} d\lambda = \int_0^\infty (J_{scE\lambda} + J_{scB\lambda}) d\lambda \tag{6.6}$$

6.1.5 Quantum Efficiency (QE)

It corresponds to the number of electrons produced in the external circuit to the solar cell by each photon of the incident spectrum.

External and internal quantum efficiencies are defined using Eqs. (6.7) and (6.8):

$$\mathrm{IQE} = \frac{J_{sc\lambda}}{q\Phi_0(1 - R)} \tag{6.7}$$

$$\mathrm{EQE} = \frac{J_{sc\lambda}}{q\Phi_0} \tag{6.8}$$

6.1.6 Spectral Response

It is defined as the ratio between the short-circuit spectral current density and the spectral radiation. Also the internal and external spectral response is defined using Eqs. (6.9) and (6.10):

$$\mathrm{ISR} = 0.808(\mathrm{IQE})\lambda \tag{6.9}$$

$$\mathrm{ESR} = 0.808(\mathrm{EQE})\lambda \tag{6.10}$$

6.2 Solar Cell Application

In order to apply the described model, in LabVIEW the virtual instrument called "photoncell.vi" was developed whose front panel is shown in Fig. 6.2.

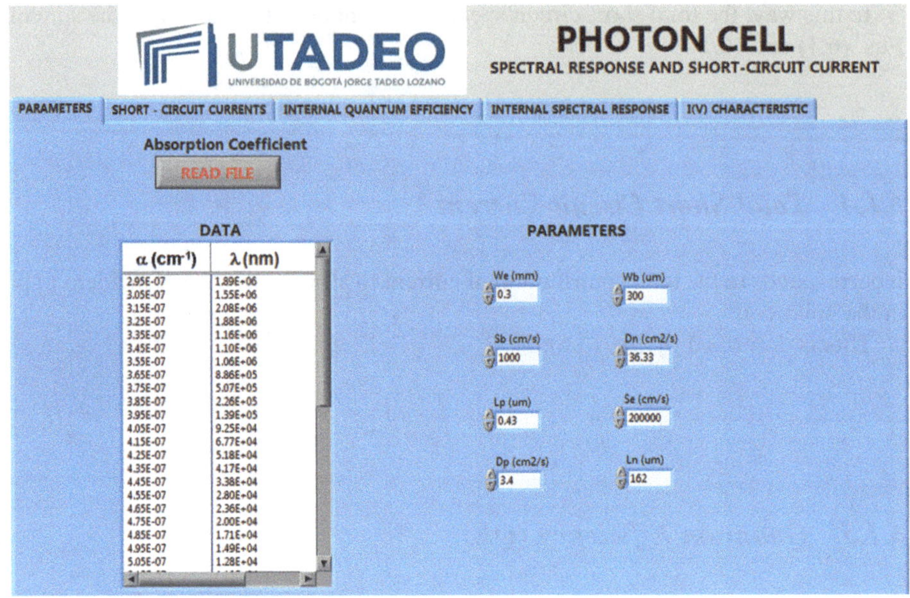

Fig. 6.2 Front panel of the "Photon Cell" program developed

The program has five selection menus:

- Parameters.
- Short-circuit currents.
- Internal quantum efficiency.
- Internal spectral response.
- $I(V)$ characteristic.

In the "Parameters" menu, the user must initially load the file containing the values of the absorption coefficient as a function of the wavelength of the solar cell to be analyzed. Then, the data related to the thickness and the other parameters indicated in Table 6.1 must be entered.

The analysis was performed considering a solar cell of silicon with the dimensions and characteristics given below [14]:

Emitter thickness: $W_e = 0.3$ mm, base thickness, $W_b = 300$ μm, $L_p = 0.43$ μm, $S_e = 2 \times 10^5$ cm/s, $D_p = 3.4$ cm^2/s, $L_n = 162$ μm, $S_b = 1000$ cm/s, $D_n = 36.33$ cm2/s.

Figure 6.3 shows the short-circuit spectral current densities for the base and the emitter of the solar cell; as well as the total current density; obtained from the data described in the previous paragraph.

An emitter short-circuit spectral current density of 24.3 mA/cm^2 (maximum value) and 61.6 mA/cm^2 (maximum value) was obtained for the base.

The program allows to store the information in. xls format files for later analysis and to create an html type report.

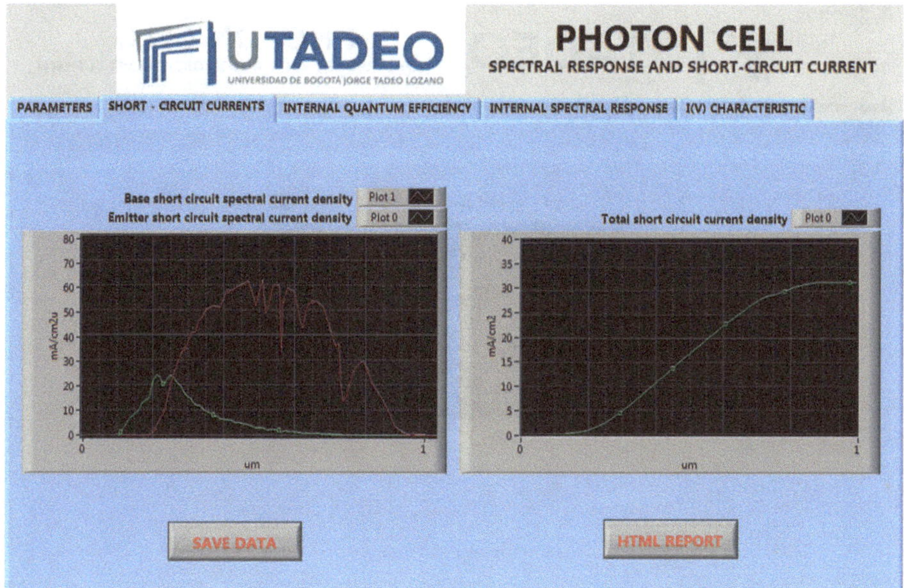

Fig. 6.3 Spectral short-circuit current densities of the emitter (*green-left*) and base (*red-left*). Total short-circuit current density (*right*)

As expected, the spectral short-circuit current density at the base of the photovoltaic device is greater than that at the emitter, precisely 40 mA/cm^2 taking into account that in the emitter there are losses in the generation of carriers associated with the surface recombination processes of the material. Treatments associated with the antireflective handling of the solar radiation in the layer of the emitter, would impact beneficially the collection of carriers that would allow to increase the response of the total current density.

Figure 6.4 shows the internal quantum efficiency of the solar cell.

The internal quantum efficiency reached 97% for the 0.65 μm incident wavelength.

The internal quantum efficiency tends to decrease significantly from 1.4 μm and this may be due to the increase of the light reflected by incident photon for each wavelength from that value (1.4 μm).

Again, the implementation of an antireflective layer could increase the amount of light reaching the base and generate more electron–hole pairs and that the former can effectively transition to the conduction band.

The internal spectral response is shown in Fig. 6.5.

The internal spectral response recorded a maximum value of 610 mA/W in the 1.45 μm.

It can be seen that for a wavelength of about 0.30 μm, the cell has an internal spectral response of approximately 180 mA/W; while at the same wavelength the internal quantum efficiency was about 70% (Fig. 6.4). This behavior evidences the

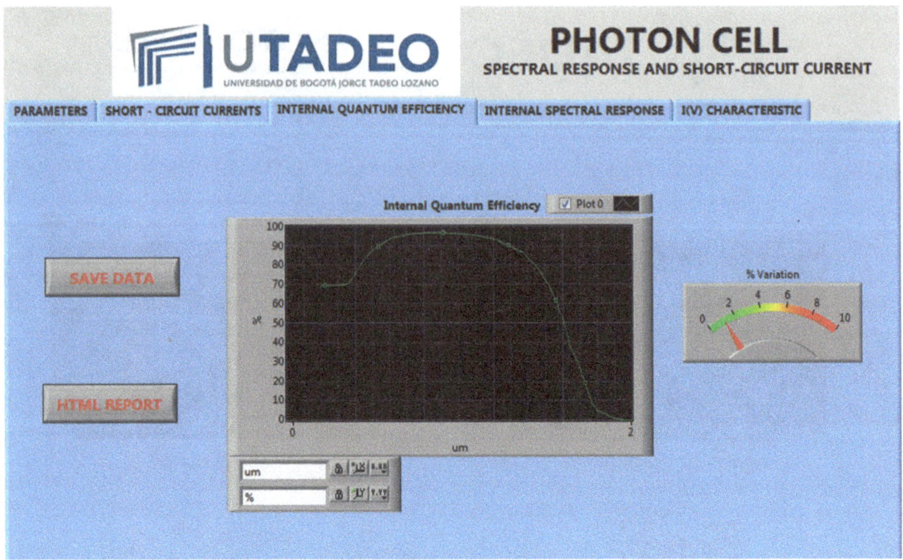

Fig. 6.4 Response of the internal quantum efficiency of the analyzed solar cell

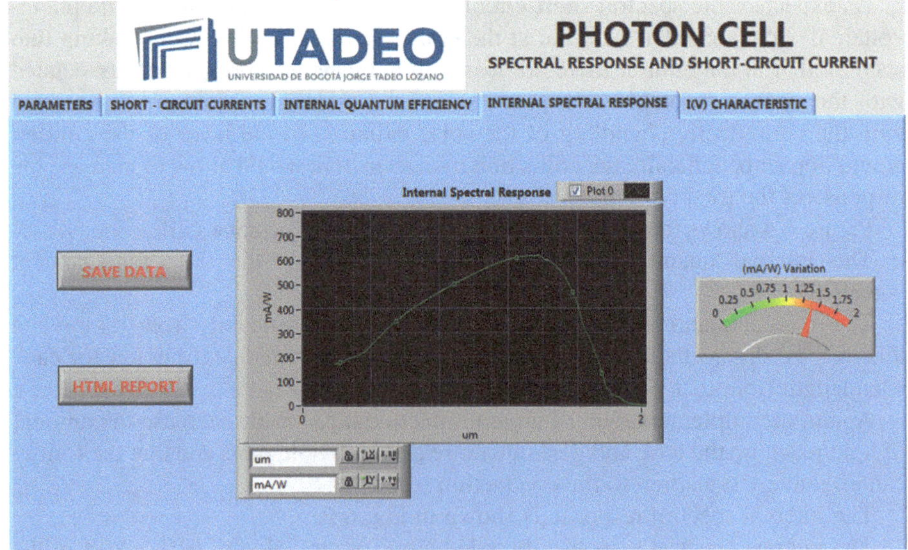

Fig. 6.5 Internal spectral response of the analyzed solar cell

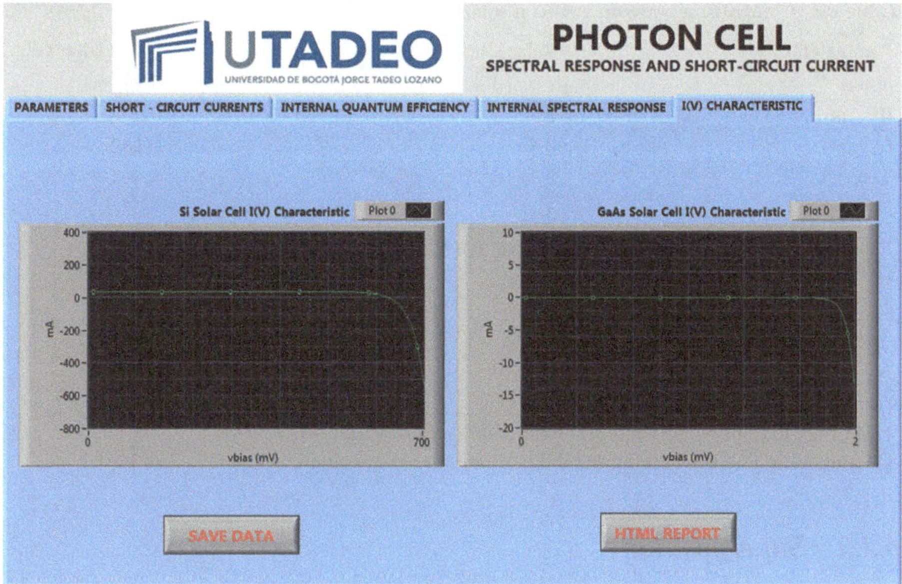

Fig. 6.6 Calculation of the characteristic I (V) for a solar cell of Si and another of GaAs

beginning of the collection of majority carriers that will define the maximum power point of the device (Fig. 6.6) at 620 mV.

6.2.1 Solar Cell's Current in the Dark

The *I–V* curve in the dark of a solar cell is identical to that of a conventional diode and is described in Eq. (6.11):

$$\mathrm{Id} = \mathrm{Is}\left(e^{\left(\frac{qV}{KT}\right)} - 1\right) \tag{6.11}$$

6.2.2 Effects of Solar Cell Material

As mentioned above, several important magnitudes in the photovoltaic conversion depend on the properties of the semiconductor material.

Table 6.2 presents the main parameters of a solar cell of silicon and of a solar cell of GaAs [14].

Table 6.2 Comparison between a silicon solar cell and a GaAs solar cell

Parameter	Solar silicon cell[b]	GaAs solar cell[a]
n_i (cm$^{-3)}$	1×10^{10}	9.27×10^5
W_e (μm)	0.3	0.2
L_p (μm)	0.43	0.132
S_e (cm/s)	2×10^5	5×10^3
D_p (cm^2/V)	3.4	7.67
W_b (μm)	300	3.8
L_n (μm)	162	1.51
S_b (cm/s)	1×10^3	5×10^3
D_n (cm^2/V)	36.63	46.8
J_0 (A/cm^2)	1×10^{-12}	9.5×10^{-20}
J_{sc} (A/cm^2)	31.188×10^{-3}	25.53×10^{-3}

[a]From L.D. Partain, Solar Cells and their Applications, Wiley 1995, p. 16
[b]Geometry file PC1D5 Pvcell.prm, and calculations using equations in annex 2 of [14]

6.2.3 *Superposition*

The superposition is used to determine the total current of the solar cell by adding the contributions under illumination and darkness.

Once the illumination currents and the dark darkness of the solar cell have been calculated, the total current is:

$$J = J_{SC} - J_{dark} \tag{6.12}$$

Substituting the current under darkness gives the Eq. (6.12):

$$J = J_{SC} - J_0\left(e^{\frac{V}{V_T}} - 1\right) \tag{6.13}$$

6.2.4 I–V *Characteristics of a Solar Cell*

In this case we use "Photon Cell" using Eq. (6.13), for a solar cell with 1 cm^2 of area and the data presented in Table 6.2 for J_{sc} and J_o for a silicon solar cell.

Simultaneously, the *I–V* characteristic for a GaAs solar cell is calculated using Eq. (6.13) and the data in Table 6.2.

The result of the *I–V* curve can be seen in Fig. 6.6.

As can be seen by comparing the characteristics *I* (*V*) for the solar cell of Si and for the GaAs, the different materials of the solar cells and their geometries produce different levels of current and voltage. In particular, it may be noted that the voltage difference is the most important in this case: the maximum GaAs solar cell voltage value is twice that of the silicon solar cell, which can be attributed in large part to the great difference in the values of Jo [14].

A tool was developed using virtual instrumentation that allows, after measuring the absorption spectrum; calculate the parameters that characterize a solar cell. It is possible to represent the behavior of the internal quantum efficiency of a solar cell with an equivalent mono crystalline model that considers effective parameters.

The diffusion length is a parameter directly related to the internal quantum efficiency of the solar cell and provides information on the transport processes as well as the phenomenon of carrier recombination. In the model used, it is considered that the collection efficiency of carriers along the space charge region remains constant. The internal quantum efficiency varies with depth and can be calculated with parameters governed exclusively by absorption processes. For the calculation of the spectral response, and under short circuit conditions; the current is linear with the irradiance and is less sensitive to small changes in voltage, depending directly on the wavelength that impinges on the solar cell. It is verified that at small wavelengths, the spectral response of the solar cell decreases and it is inferred that the energy greater than the gap of the material; is actually used by the solar cell. The calculation of the described parameters allows to analyze other properties of the solar cells, useful to characterize in an integral way the performance and electrical efficiency of the photovoltaic devices; as well as to deepen in aspects that affect the processes of recombination and the electric transport.

References

1. J. Bisquert, Dilemmas of Dye sensitized solar cells. ChemPhysChem **12**, 1633–1636 (2011)
2. I. Chung, B. Lee, J. He, R.P.H. Chang, M.G. Kanatzidis, All solid state dye sensitized solar cells with high efficiency. Nature **485** (2012)
3. A. Hagfeldt, G. Boschloo, L.C. Sun, L. Kloo, H. Pettersson, Dye-sensitized solar cells. Chem. Rev. **110**, 6595–6663 (2010)
4. M. Gratzel, Recent advances in sensitized mesoscopic solar cells. Acc. Chem. Res. **42**, 1788–1798 (2009)
5. H. Yang, H. Wang, M. Wang, Investigation of open-circuit voltage in solar grade silicon solar cells from a metallurgical process route and cell's defect. Clean Techn. Environ. Policy **15**, 111–116 (2013)
6. A.A. Istratov, T. Buonassisi, R.J. McDonald, et al., Metal content of multicrystalline silicon for solar cells and its impact on minority carrier diffusion length. J. Appl. Phys. **94**, 6552–6557 (2003)
7. B. Andó, S. Graziani, N. Pitrone, Stand-alone laboratory sessions in sensors and signal process- ing. IEEE Trans. Educ. **47**, 4–9 (2004)
8. M. Nagistris, A Matlab-based virtual laboratory for teaching introductory quasi-stationary electro-magnetics. IEEE Trans. Educ. **48**, 81–88 (2005)
9. T.W. Gedra, A. Seungwon, Q.H. Arsalan, S. Ray, Unified power engineering laboratory for electromechanical energy conversion, power electronics and power systems. IEEE Trans. Power Systems **19**, 112–119 (2004)
10. Desarrollo de celdas solares basadas en CuInSe2 usando nuevos materiales buffer en su estructura. Clara Lilia Calderon Triana. Doctoral thesis. Universidad Nacional de Colombia (2003)
11. A. Luque, S. Hegedus (eds.), *Handbook of Photovoltaic Science and Engineering* (John Wiley & Sons, Ltd, New York, 2003.) ISBN: 0-471-49196-9

12. Síntesis y Caracterización de Nuevos Materiales No Tóxicos Empleados como Capa Buffer y Capa Absorbente en la Fabricación de Celdas Solares. Mónica Andrea Botero Londoño. Doctoral thesis. Universidad Nacional de Colombia. Bogotá (2008)
13. A. L. Fahrenbruch, R. H. Bibe (eds.), *Fundamentals of Solar Cells. Photovoltaic Solar Energy Conversión* (Academia Press, Inc., New York, 1983.) ISBN: 0-12-247680-8
14. L. Castañer, S. Silvestre, *Modelling Photovoltaic Systems Using PSpice* (Ed. Wiley, New York, 2002)

Chapter 7
PV Generator Characterization

The process of a PV cell behavior's simulation and characterization is based on its equivalence with an electric circuit and its subsequent mathematical representation. The most important characteristic of a solar cell for its electrical simulation is the dependence of the current that flows through it in function of the applied voltage. For this, constant values of incident radiation intensity and temperature are maintained.

The performance curves of the photovoltaic panel *I–V* and *P–V*, Fig. 7.1, show the possible combinations of current and voltage for a photovoltaic device under certain ambient conditions (incident solar radiation and ambient temperature). The particular point of current and voltage in which the photovoltaic device will work will be determined by the load to which it is connected.

7.1 Parameters of the *I–V* Characteristic

- *Short circuit current (Isc)*: it is the maximum current that will produce the device under defined conditions of lighting and temperature, corresponding to a voltage equal to zero.
- *Open circuit voltage (Voc)*: The maximum voltage of the device under defined lighting and temperature conditions, corresponding to a current equal to zero.
- *Maximum power (Pmax)*: It is the maximum power that will produce the device under certain conditions of lighting and temperature, corresponding to the maximum parameter *I–V*.
- *Current at maximum power point (Impp)*: It is the current value for Pmax under certain lighting and temperature conditions.
- *Voltage at maximum power point (Vmpp)*: It is the voltage value for Pmax under certain conditions of illumination and temperature.

A.J. Aristizábal Cardona et al., *Building-Integrated Photovoltaic Systems (BIPVS)*,
https://doi.org/10.1007/978-3-319-71931-3_7

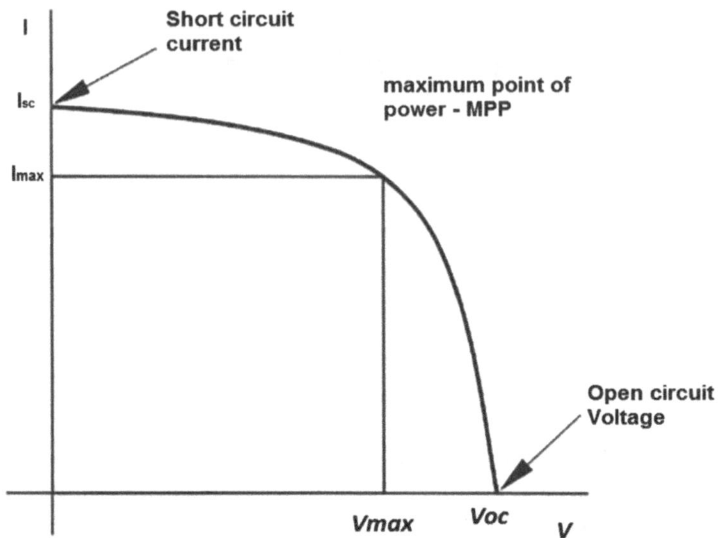

Fig. 7.1 Characteristic curve of a PV panel

7.2 Electrical Model of the Photovoltaic Cell

Figure 7.2 shows the model of the equivalent electric circuit of a solar photovoltaic cell. The current source Iph represents the photocurrent generated, the diode D represents the P/N junction, Rs represents the serial resistance of the device associated with the resistance of the materials and the electrical contacts, and Rsh represents the parallel resistance of the device related to the current leakage in the device volume [1].

Typically the photoelectric cells are presented in series interconnected cell arrays to increase the output voltage to a desired value and in parallel to increase the electrical current that the device can supply according to the power requirements [2]. Each solar cell behaves in the dark in a manner similar to a rectifier P/N diode and under the light incidence it generates an electric photocurrent.

Next, the description of the mathematical model for the equivalent electric circuit of a solar photovoltaic cell [3–5].

Applying Kirchhoff's current law, it is found that the current in the terminals of the photovoltaic panel is:

$$\mathrm{Ipv} = \mathrm{Iph} - \mathrm{ID} - \mathrm{IRsh} \tag{7.1}$$

Where Ipv is the current generated by the system, Iph is the photocurrent generated, ID is the current that flows through the diode, and IRsh is the current that flows through the parallel resistance.

Equation 7.2 refers to the photocurrent generated Iph that depends on the radiation and the standard temperature [4], G is the radiation at the current site,

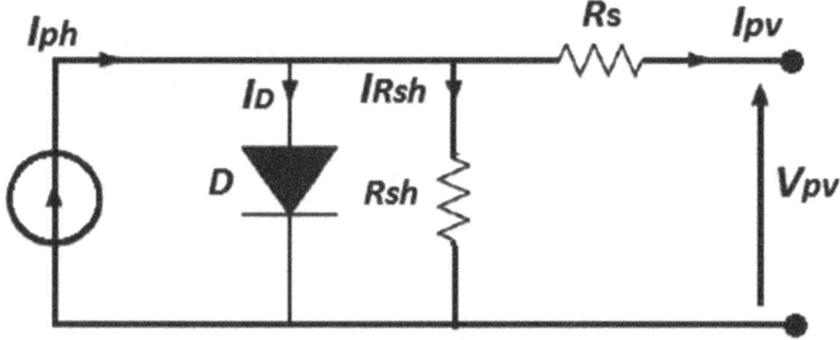

Fig. 7.2 Equivalent electrical model of a PV cell. Source [1]

T and Tr are the present temperature of the site and the reference temperature, Δi is the temperature coefficient for the current, and Isc is the short-circuit current of the cell at standard temperature.

$$\mathrm{Iph} = \frac{G}{1000}[\mathrm{Isc} + \Delta i(T - \mathrm{Tr})] \tag{7.2}$$

The saturation current of the ID diode is obtained by the Shockley equation.

$$\mathrm{ID} = \mathrm{Isat}\left[\exp\left(\frac{\mathrm{VD}}{\mathrm{Vt}}\right)\right] - 1 \tag{7.3}$$

Where Isat corresponds to the diode's saturation current, VD is the diode voltage, and Vt is the diode thermodynamic voltage.

Applying Kirchhoff's voltage law, then:

$$\mathrm{VD} = \mathrm{Vpv} + \mathrm{Rs} * \mathrm{Ipv} \tag{7.4}$$

Replacing VD and isolating IRsh then:

$$\mathrm{IRsh} = \frac{\mathrm{Vpv} + \mathrm{Rs} * \mathrm{Ipv}}{\mathrm{Rsh}} \tag{7.5}$$

It is also possible:

$$\mathrm{Vt} = \frac{A * K * \mathrm{Tc}}{q} \tag{7.6}$$

where, k corresponds to the Boltzmann constant, Tc refers to the temperature at the P/N junction, q is the value of the charge of the electron, and A corresponds to the diode ideality factor.

The electrical characteristics of the photovoltaic panel are given in terms of current and output voltage (Ipv − Vpv) as follows:

$$\mathrm{Ipv} = \mathrm{Iph} - \mathrm{Isat}\left[\exp\left(\frac{\mathrm{VD}}{\mathrm{Vt}}\right) - 1\right] - \frac{\mathrm{Vpv} + \mathrm{Rs}*\mathrm{Ipv}}{\mathrm{Rsh}} \tag{7.7}$$

The Isat parameter needs to be evaluated as open circuit, where

$$\mathrm{Ipv} = 0 \tag{7.8}$$

$$\mathrm{Vpv} = \mathrm{Voc} \tag{7.9}$$

In these conditions, it is rewritten as follows:

$$0 = \mathrm{Iph} - \mathrm{Isat}\left[\exp\left(\frac{\mathrm{Voc}}{\mathrm{Vt}}\right) - 1\right] - \frac{\mathrm{Voc}}{\mathrm{Rsh}} \tag{7.10}$$

And simplifying:

$$\mathrm{Isat} = \frac{\mathrm{Iph} - \frac{\mathrm{Voc}}{\mathrm{Rsh}}}{\exp\left(\frac{\mathrm{Voc}}{\mathrm{Vt}}\right) - 1} \tag{7.11}$$

The reverse saturation current also depends on the temperature according to [4].

$$\mathrm{Isat} = \mathrm{Isatr}*\left(\frac{\mathrm{Tc}}{\mathrm{Tr}}\right)^3 *\exp\left[\frac{q*\mathrm{Eq}}{K*A}\right]\left[\frac{1}{\mathrm{Tr}} - \frac{1}{\mathrm{TC}}\right] \tag{7.12}$$

$$\mathrm{Tc} = \mathrm{Ta} + (0.2*G) \tag{7.13}$$

where Isatr is the reference saturation current, Tc is the cell temperature, Tr is the reference temperature, Eq is the gap's band energy, q is the electron charge, and K is the Boltzmann constant, A Corresponds to the diode ideality factor, Ta is the ambient temperature, and G is the radiation.

And Short Circuit:

$$\mathrm{Ipv} = \mathrm{Icc} \tag{7.14}$$

$$\mathrm{Vpv} = 0 \tag{7.15}$$

In these conditions:

$$\mathrm{Icc} = \mathrm{Iph} - \frac{\mathrm{Iph} - \frac{\mathrm{Voc}}{\mathrm{Rsh}}}{\exp\left(\frac{\mathrm{Voc}}{\mathrm{Vt}}\right) - 1} - \frac{\mathrm{Icc}*\mathrm{Rs}}{\mathrm{Rsh}} \tag{7.16}$$

Rewrites:

$$\mathrm{Iph} = \frac{\mathrm{Icc}\left(1 - \frac{\mathrm{Rs}}{\mathrm{Rsh}}\right) - \mathrm{Voc}\left(\frac{\exp\left(\frac{\mathrm{Icc}*\mathrm{Rs}}{\mathrm{Vt}}\right)-1}{\exp\left(\frac{\mathrm{Voc}}{\mathrm{Vt}}\right)-1}\right)}{1 - \left(\frac{\exp\left(\frac{\mathrm{Icc}*\mathrm{Rs}}{\mathrm{Vt}}\right)-1}{\exp\left(\frac{\mathrm{Voc}}{\mathrm{Vt}}\right)-1}\right)} \tag{7.17}$$

Rs is the resistance in series and Rsh is the resistance in parallel as shown in [6].

$$\mathrm{Rs} = \left[1 - \frac{\mathrm{FF}}{\mathrm{FFo}}\right]\left(\frac{\mathrm{Voc,STC}}{\mathrm{Isc,STC}}\right) \tag{7.18}$$

$$Rsh = \frac{FFo(Vo + 0.7)}{\left(1 - \frac{FF}{FFo}\right)Vo}\left(\frac{Voc,STC}{Isc,STC}\right) \tag{7.19}$$

where FF is the Fill Factor and FFo is the ideal Fill Factor according to [6]; when Rs is = 0 it is obtain.

$$\mathrm{FF} = \frac{\mathrm{Pm}}{\mathrm{Voc}*\mathrm{Isc}} = \frac{\mathrm{Vm}*\mathrm{Im}}{\mathrm{Voc}*\mathrm{Isc}} \tag{7.20}$$

$$\mathrm{FFo} = \frac{\mathrm{Vo} - \mathrm{Ln}(\mathrm{Vo} + 0.72)}{\mathrm{Vo}};\ \mathrm{Donde} \quad \mathrm{Vo} = \frac{\mathrm{Voc}}{\mathrm{KTc}} \tag{7.21}$$

Replacing Eqs. (7.11) and (7.17) in Eq. (7.7) gives the following Eqs. (7.22) and (7.23):

$$\mathrm{Ipv} = \mathrm{Iph} - \frac{\mathrm{Iph} - \frac{\mathrm{Voc}}{\mathrm{Rsh}}}{\exp\left(\frac{\mathrm{Voc}}{\mathrm{Vt}}\right) - 1}\left(\exp^{\frac{\mathrm{Vpv}+\mathrm{Ipv}*\mathrm{Rs}}{\mathrm{VT}}} - 1\right) - \frac{\mathrm{Vpv} + \mathrm{Ipv}*\mathrm{Rs}}{\mathrm{Rsh}} \tag{7.22}$$

$$\mathrm{Ipv} = \frac{\mathrm{Icc}\left(1 + \frac{\mathrm{Rs}}{\mathrm{Rsh}}\right) - \mathrm{Voc}\left(\frac{\exp\left(\frac{\mathrm{Icc}*\mathrm{Rs}}{\mathrm{Vt}}\right)-1}{\exp\left(\frac{\mathrm{Voc}}{\mathrm{Vt}}\right)-1}\right)}{1 - \left(\frac{\exp\left(\frac{\mathrm{Icc}*\mathrm{Rs}}{\mathrm{Vt}}\right)-1}{\exp\left(\frac{\mathrm{Voc}}{\mathrm{Vt}}\right)-1}\right)} - \left(\frac{\mathrm{Icc}\left(1 + \frac{\mathrm{Rs}}{\mathrm{Rsh}}\right) - \mathrm{Voc}\left(\frac{\exp\left(\frac{\mathrm{Icc}*\mathrm{Rs}}{\mathrm{Vt}}\right)-1}{\exp\left(\frac{\mathrm{Voc}}{\mathrm{Vt}}\right)-1}\right)}{\left(\exp\left(\frac{\mathrm{Voc}}{\mathrm{Vt}}\right) - 1\right) - \left(\exp\left(\frac{\mathrm{Icc}*\mathrm{Rs}}{\mathrm{Vt}}\right) - 1\right)} - \frac{\mathrm{Voc}}{\left(\exp\left(\frac{\mathrm{Voc}}{\mathrm{Vt}}\right) - 1\right)*\mathrm{Rsh}}\right)$$

$$\left(\exp^{\frac{\mathrm{Vpv}+\mathrm{Ipv}*\mathrm{Rs}}{\mathrm{VT}}} - 1\right) - \frac{\mathrm{Vpv} + \mathrm{Ipv}*\mathrm{Rs}}{\mathrm{Rsh}} \tag{7.23}$$

The cell temperature (Tc) is determined by taking into account the ambient temperature (Ta) at the moment of operation and the NOCT (nominal cell operating temperature) delivered by the manufacturer.

$$\mathrm{Tsc} = \mathrm{Ta} + \frac{\mathrm{NOCT} - 20}{800} G\left(\mathrm{W/m^2}\right) \tag{7.24}$$

The following equations correspond to the parallel and serial impedances of the photovoltaic panel where Np is the number of cells in parallel, Ns is the number of cells in series, Rp is the resistance in parallel, and Rs is the series resistance.

$$\mathrm{Rsht} = \left(\frac{\mathrm{Np}}{\mathrm{Ns}}\right)\mathrm{Rp} \tag{7.25}$$

$$\mathrm{Rst} = \left(\frac{\mathrm{Ns}}{\mathrm{Np}}\right)\mathrm{Rs} \tag{7.26}$$

7.3 Simulation and Characterization for Different Levels of Irradiance

After performing the mathematical analysis of the photovoltaic panel's electrical model, it proceeds to evaluate the characteristics of one of the photovoltaic panels installed at the Universidad de Bogota Jorge Tadeo Lozano. Using and adjusting the code developed in Matlab [4] where the *I–V* and *P–V* performance curves of the PV panel are obtained by entering the ambient temperature and the irradiance as initial conditions.

To verify the implemented model, these parameters are entered as initial condition to plot the characteristics of the solar panel according to the variables that the manufacturer delivers and that are conditioned to the source code (simulated characteristic). Subsequently, into the source code it is entered a vector with the parameters of voltage, current and power of the characteristic curve that the manufacturer delivers in the specification sheet and in this way to be able to graph and compare the similarity of the synchronism of the curves (*I–V* And *P–V*,). Finally it is determine the absolute error and the relative error that exist between the two curves (simulated curve Vs fabric curve).

The analyzed figures correspond to the typical curves of IV and PV, under Standard Test Conditions (STC: irradiance 1000 W/m^2, ambient temperature of 25 °C and air mass AM 1.5) and under conditions of Nominal Operating cell Temperature (NOCT) of the photovoltaic module TSM-PA05.08 Trina Solar according to IEC 60904-3 [7].

The analysis will focus on the most important areas of the IV and PV characteristic curve, such as the Maximum Power Point (MPP), as this area generates the most power and performance extraction of the photovoltaic module [8].

As seen in Figs. 7.3 and 7.4 the analysis and study of the MPP point is based on determining the result of the absolute error and the relative error that exists between the experimental curve delivered by the manufacturer and the simulated curve theoretically, all this for the purpose to validate the proposed model.

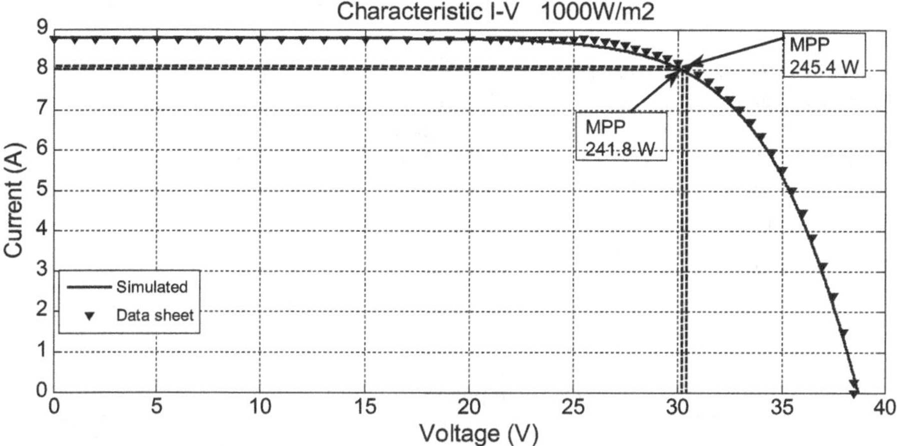

Fig. 7.3 Typical comparison of the *I–V* characteristic, specification sheet, and theoretical simulation of the TSM-PA05.08 module under parameters of irradiance of 1000 W/m^2 and temperature of 25 °C

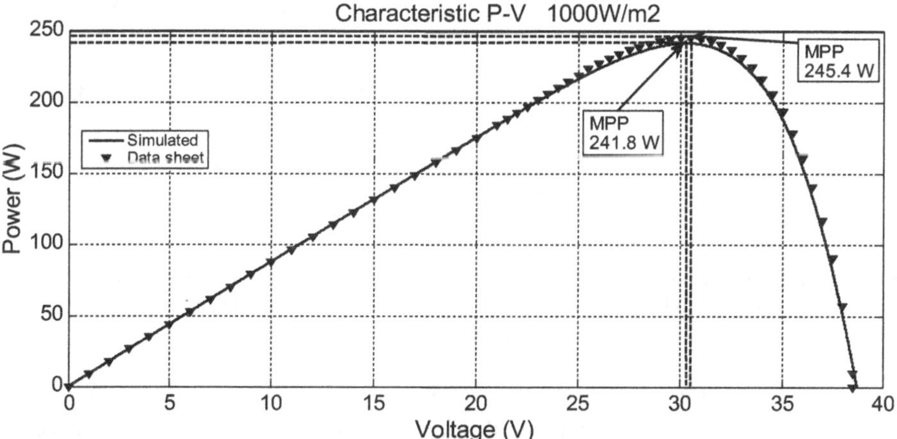

Fig. 7.4 Typical comparison of the characteristic *P–V*, specification sheet, and theoretical simulation of the TSM-PA05.08 module under irradiance parameters of 1000 W/m^2 and temperature of 25 °C

- Characteristic curves (*I–V* and *P–V*) for an Irradiance of 1000 W/m^2

As shown in Table 7.1, the maximum point of power between the two curves (MPP) of Figs. 7.3 and 7.4 shows a relative error of 1.5%. This slight separation between the curves shows the good performance of the proposed model since the rest of the graph remains in synchronism.

- Characteristic curve (*I–V* and *P–V*) for an Irradiance of 800 W/m^2

Table 7.1 Absolute error and relative error for MPP at different levels of irradiance and temperature of 25 °C

Irradiance	Power simulated	Power data sheet	Power absolute error	Power relative error
W/m^2	MPPT	MPPT	εa = [Real value − Approximate value]	εr = (εa/Real value) * 100
1000	241.8	245.4	3.6	1.5
800	191.1	194.9	3.8	1.9
600	140.6	144.3	3.7	2.6
400	90.9	94.8	3.9	4.2

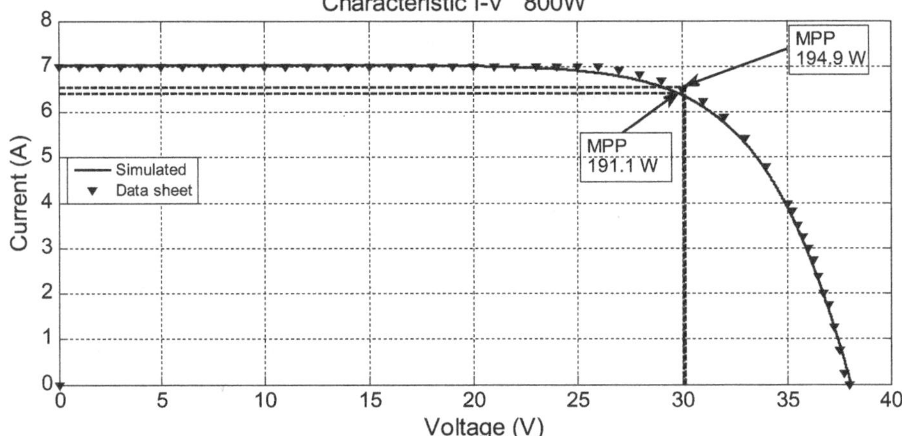

Fig. 7.5 Typical comparison of the *I–V* characteristic, specification sheet, and theoretical simulation of the TSM-PA05.08 module under parameters of irradiance of 800 W/m^2 and temperature of 25 °C

Likewise, the Figs. 7.5 and 7.6 corresponding to the characteristic curves *I–V* and *P–V* are analyzed and compared, under ambient conditions of irradiance of 800 W/m^2 and ambient temperature of 25 °C. The analysis and result of the absolute error and relative error of the MPP are observed in Table 7.1, where again a slight separation between the curves at this point of 1.95% is observed, compared to the rest of the graph that remains in synchronism.

- Characteristic curve (*I–V* and *P–V*) for Irradiance of 600 W/m^2

The same analysis is made to results in Figs. 7.7 and 7.8 corresponding to the typical *I–V* and *P-V* curves, under ambient irradiance conditions of 600 W/m^2 and 25 °C ambient temperature of the TSM-PA05.08 photovoltaic module. The analysis of the maximum power point of Figs. 7.7 and 7.8 shows the absolute and relative error of the point of greatest energy extraction, as shown in Table 7.1, with a separation between the curves of 2.6%.

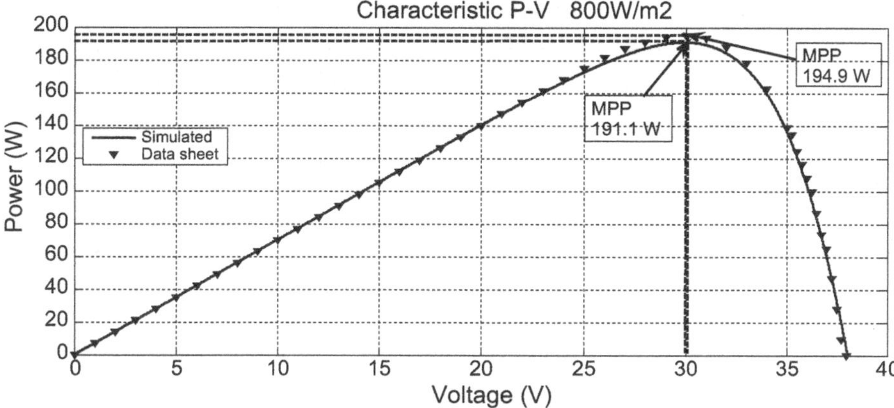

Fig. 7.6 Typical comparison of the characteristic *P–V*, specification sheet, and theoretical simulation of the TSM-PA05.08 module under parameters of irradiance of 800 W/m^2 and temperature of 25 °C

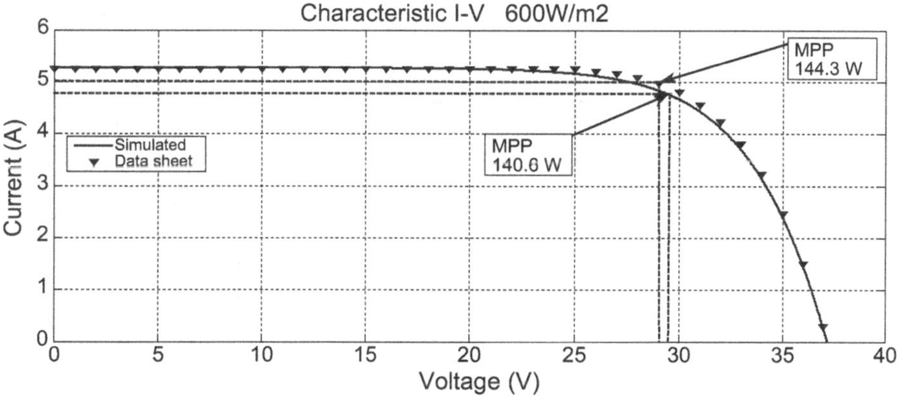

Fig. 7.7 Typical comparison of the *I–V* characteristic, specification sheet, and theoretical simulation of the TSM-PA05.08 module under irradiance parameters of 600 W/m^2 and temperature of 25 °C

- Characteristic curve (*I–V* and *P–V*) for an Irradiance of 400 W/m^2

The analysis is also performed in results on Figs. 7.9 and 7.10 corresponding to the typical curves of *I–V* and the characteristic curve *P-V*, under ambient irradiance conditions of 400 W/m^2 and ambient temperature of 25 °C. The MPP area analysis of Figs. 7.9 and 7.10 can be seen in Table 7.1.

- Relative Error of the Curves (*I–V* and *P–V*) in the MPP for Different Levels of Irradiance

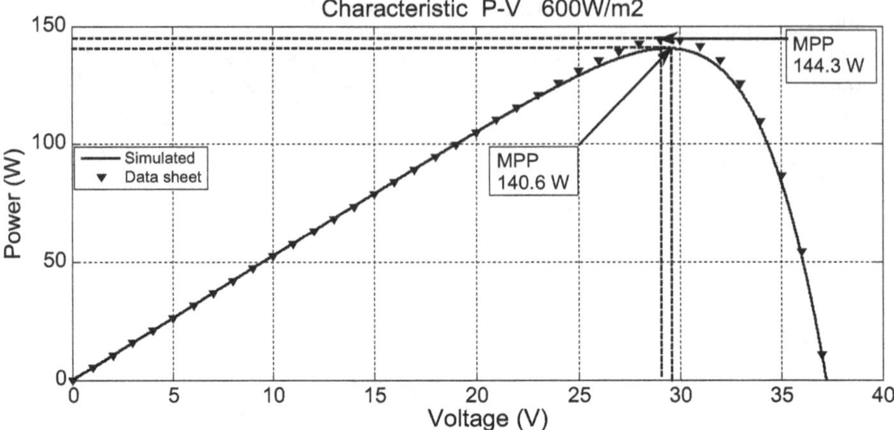

Fig. 7.8 Typical comparison of the characteristic *P–V*, specification sheet, and theoretical simulation of the TSM-PA05.08 module under irradiance parameters of 600 W/m^2 and temperature of 25 °C

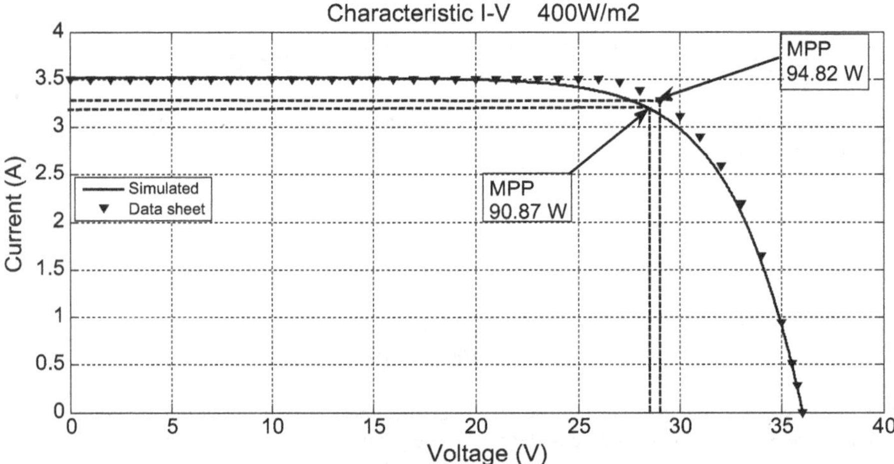

Fig. 7.9 Typical comparison of the *I–V* characteristic, specification sheet, and theoretical simulation of the TSM-PA05.08 module under irradiance parameters of 400 W/m^2 and temperature of 25 °C

Figures 7.11 and 7.12 show the characteristic curves of *I–V* and *P–V* for different levels of irradiance with simulation performed in Matlab. The plotted curves refer to the curves of the technical specifications sheet supplied by the manufacturer and the simulated curves with the theoretical parameters of the TSM-PA05.08 photovoltaic module.

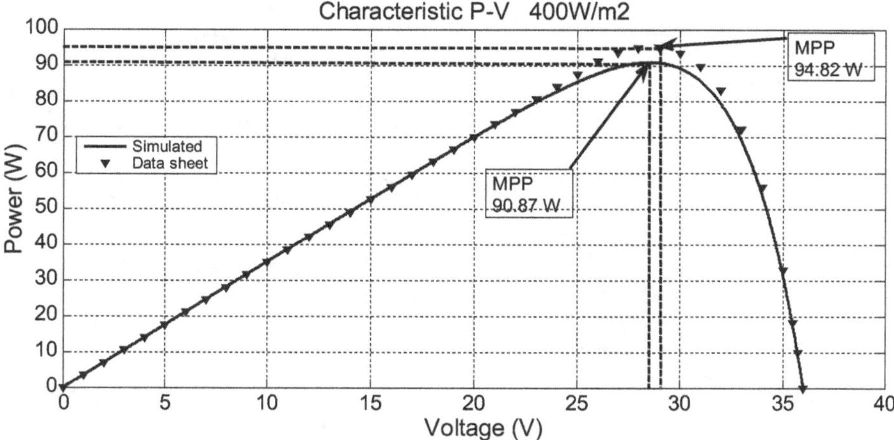

Fig. 7.10 Typical comparison of the characteristic *P–V*, specification sheet, and theoretical simulation of the TSM-PA05.08 module under parameters of irradiance of 400 W/m^2 and temperature of 25 °C

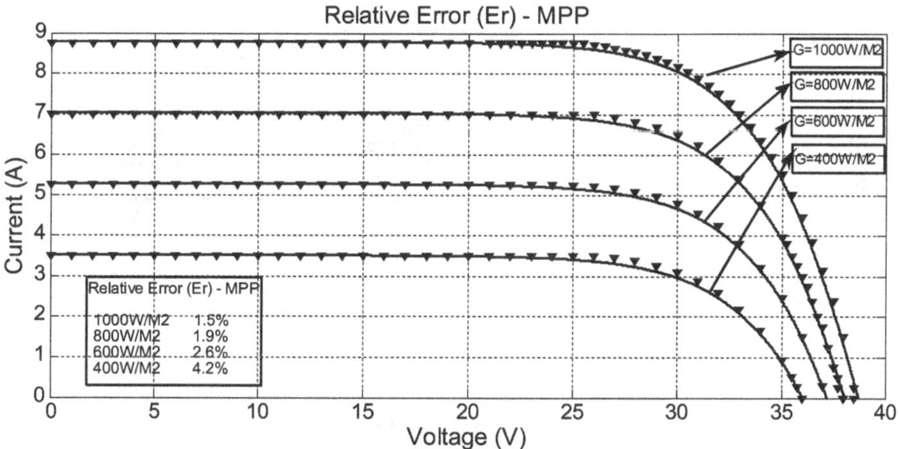

Fig. 7.11 Relative Error of the *I–V* Curves of the MPP, specification sheet, and theoretical simulation of the TSM-PA05.08 module

The analysis of the relative error of the maximum power point of each curve for different levels of irradiance is summarized in Table 7.1.

- (*I–V* and *P–V*) curves for Different Temperature Levels

Figures 7.13 and 7.14 show the characteristic curves of *I–V* and *P–V* for different temperature levels with constant irradiance at 1000 W/m^2. It can be analyzed that the voltage of the solar module is affected as the temperature increases, or that the

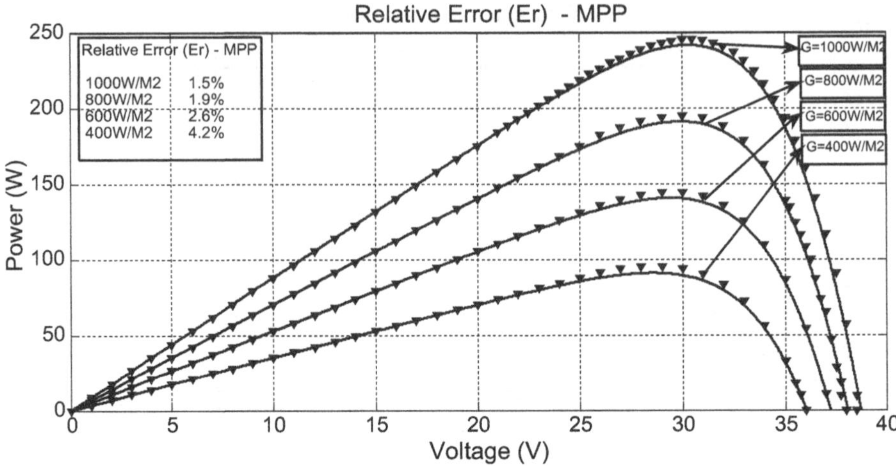

Fig. 7.12 Relative Error of the *P–V* Curves of the MPP, specification sheet, and theoretical simulation of the TSM-PA05.08 module

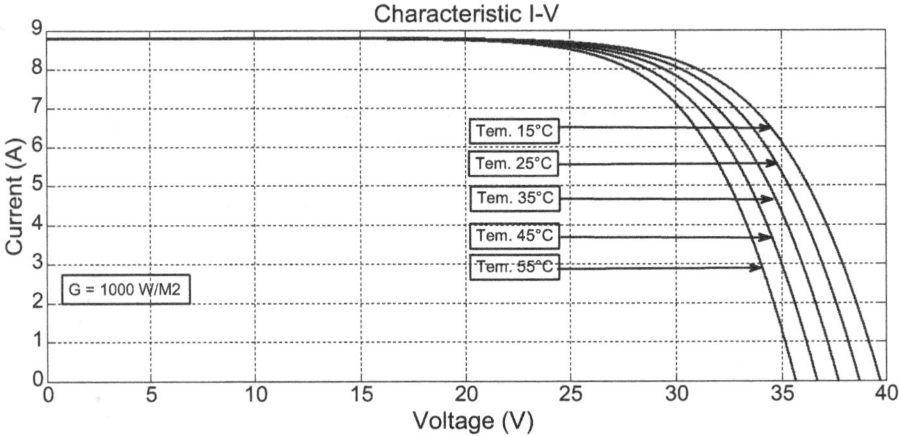

Fig. 7.13 *I–V* curves, for different temperature scales and constant irradiance at 1000 W/m^2 of module TSM-PA05.08

voltage decreases when subjected to high temperatures, directly affecting the performance of the photovoltaic module.

- Curves (*I–V* and *P–V*) of the 6 KW Photovoltaic Generator

Figures 7.15 and 7.16 show the *I–V* and *P–V* characteristic curves of the 6 KW photovoltaic generator implemented at the UJTL facilities, with a temperature of 25 °C and with an irradiance of 1000 W/m^2. It can be analyzed that the open circuit voltage (Voc) reaches 456 V, according to the serial arrangement of the 12 modules

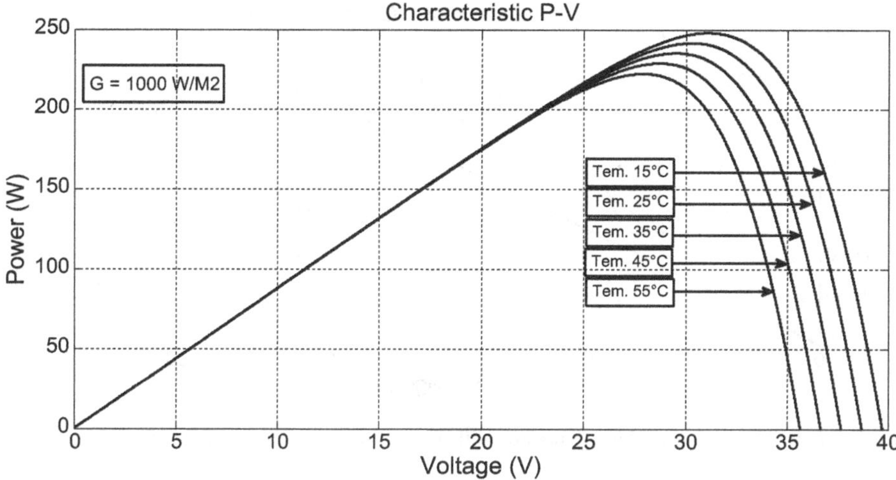

Fig. 7.14 Curves *P–V*, for different temperature scales and constant irradiance at 1000 W/m^2 of module TSM-PA05.08

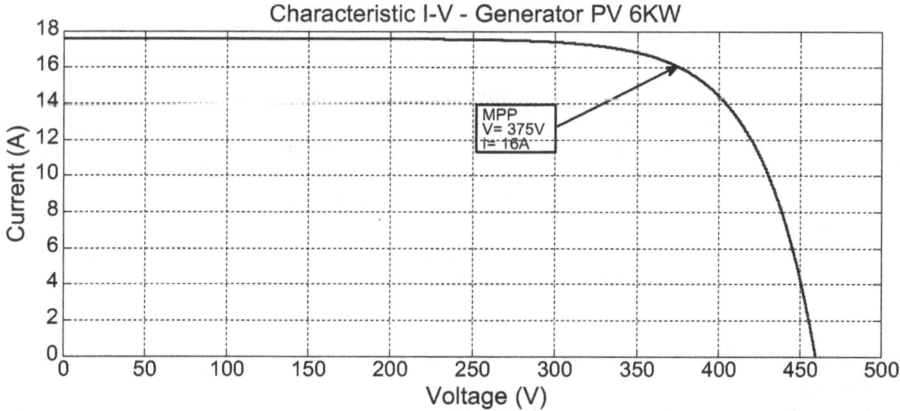

Fig. 7.15 *I–V* curves, 6 KW PV generator with 1000 W/m^2 irradiance and 25 °C temperature

of the PV array, the same thing happens with the short circuit current (Icc) reaching 17.6 A According to the arrangement of two photovoltaic arrays in parallel. It is analyzed and concluded that the model developed allows for designing, characterizing, and optimizing any type of photovoltaic generator, since the model helps to make different configurations and allows for finding the maximum power point of any PV array.

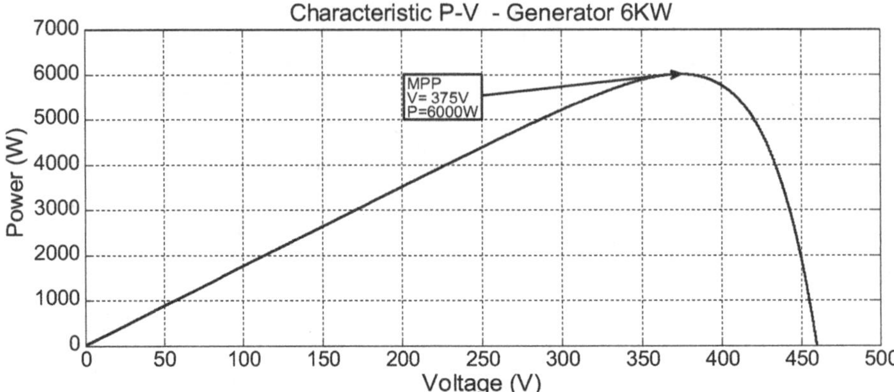

Fig. 7.16 *P–V* curves, 6 KW PV generator with 1000 W/m^2 irradiance and 25 °C temperature

Acknowledgements Special thanks to professor Alonso de Jesús Chica Leal for his guidance with the model.

References

1. T. Markvart, L. Castañer, *Practical Handbook of Photovoltaics: Fundamentals and Applications* (Elsevier, Oxford, UK, 2003)
2. J. Beltran, Prototipo fotovoltaico con seguimiento del Sol para procesos electroquímicos (2007)
3. J.A. Gow, C.D. Manning, Development of a phovovoltaic array model for use in power electronics simulation studies. IEE Proc. Elec. Power Appl. **146**(2), 193–200 (1999.) Framework
4. I. Lopez, Projeto de um Sistema Fotovoltaico Autonomo de Suprimento de Energia Usando Tecnica MPPT e Controle Digital. Universidade Federal de Minas Gerais. Belo Horizonte – Brazil (2009)
5. H.T. Duru, Sol. Energy **80**(7), 812–822 (2006)
6. M.A. Green, *Solar cells. Operating principles, technology and system applications* (Prentice-Hall, NJ, 1982)
7. IEC 60904-3. Photovoltaic devices – Part 3: Measurement principles for terrestrial photovoltaic (PV) solar devices with reference spectral irradiance data (2008)
8. A. Bayod, J. Cebollero, A novel MPPT method for PV systems with irradiance measurement, (Aug 2014)

Chapter 8
Implementation of the BIPVS Monitoring System

To analyze the BIPVS performance, a monitoring system was implemented using virtual instrumentation [1]. Signal capture is processed through a developed virtual instrument (VI), using National Instruments (NI) devices as hardware and the LabVIEW graphics programming package as software. Programs built in LabVIEW are known as Virtual Instruments (VIs), and are composed of two fundamental parts: the front panel and the block diagram.

The monitoring system has the structure shown in Fig. 8.1.

8.1 Monitoring Devices

The devices used to measure and monitor the performance variables and the power quality generated by the BIPVS are:

- Radiation sensor: The selected radiation sensor is a SP LITE 2 photodiode pyranometer from Kipp & Zonen, with a sensitivity range of 71.8 $\mu V/Wm^{-2}$, a spectral range of 400–1100 nm, and a maximum irradiation of 2000 W/m^2.
- Ambient temperature sensor: An NTC 10 KΩ thermistor was used as the temperature sensor. The variation of the electrical resistance of this device as a function of temperature was the property used to determine the temperature of the device. Temperature sensor calibration was performed through the formation of an RTerm resistance characteristic curve, by measuring the output voltage as a function of temperature T. The equation describing the calibration of the device is as follows.

$$Y = -21.08X + 76.042 \tag{8.1}$$

A.J. Aristizábal Cardona et al., *Building-Integrated Photovoltaic Systems (BIPVS)*,
https://doi.org/10.1007/978-3-319-71931-3_8

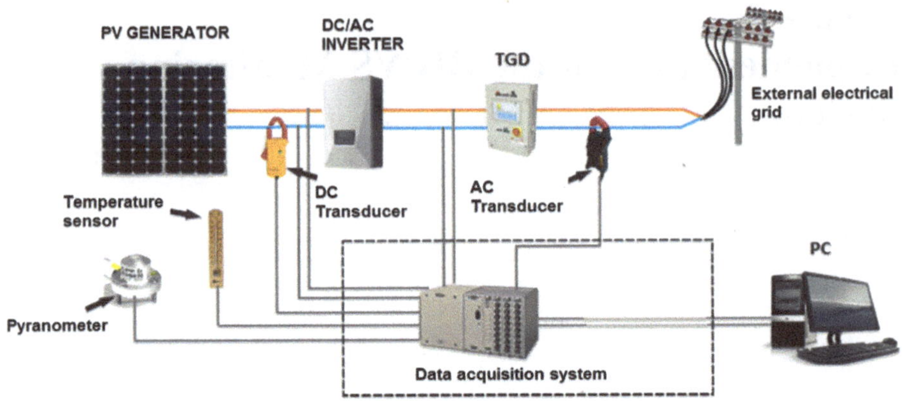

Fig. 8.1 BIPVS data acquisition and monitoring system

- Direct and Alternating Current Transducers: For the measurements of the direct current (DC) produced by the PV generator, the Fluke i410 CAT III reference clamp was used with the following technical characteristics:
  - Output signal: 1 mV/A.
  - Maximum working voltage: 600 Vrms.
  - Measurement type: Hall—Sensor.
  - Safety: CAT III—600 V.
  - Usable current range: 0.5–400 A.
  - Basic accuracy: 3.5% reading +0.5 A.
  - Usable DC frequency: 3 KHz.
- Measuring the AC alternating current at the inverter output required a Fluke i200s CAT III reference clamp (probe) with the following technical characteristics:
  - Output signals: 1 mV/A.
  - Measurement range: 0.5–200 A.
  - Usable AC frequency: 40 Hz–10 KHz
  - Basic accuracy: 1% reading +0.5 A (48–65 Hz)
  - Safety: CAT III—600 V AC.
  - Maximum current: 200 A.
  - Operating scales: 100 mV/A-20 A, 10 mV/A-200 A.

8.2 Data Acquisition System

The Data Acquisition System is in charge of the digitization of the analog signals and its characteristics establish the number of channels of signal acquisition, the speed of such acquisition, the resolution, and even the mode of acquisition of the data of the complete system.

LabVIEW Virtual Instrumentation software is in charge of the process of real-time monitoring, generation of alarms, analysis and storage of information; additionally allows interaction between the user and the monitoring system serves as interface between signal conditioning and data acquisition.

The devices selected to perform the data acquisition are:

- The NI cDAQ-9174 is a 4-slot NI CompactDAQ USB chassis designed for small and portable mixed measurement systems. Up to four C-Series I/O modules can be combined for a custom measurement system for analog input, analog output, digital I/O, and counters/timers.

Modules are available for a variety of sensor measurements including thermocouples, RTDs, extensiometric gauges, pressure and load transducers, torsion cells, accelerometers, flow meters, and microphones. NI CompactDAQ systems combine sensor measurements with voltage signals, current to create custom mixed signal systems with a single USB cable to the PC or laptop.

- The NI 9205 is a C Series module for use with the NI CompactDAQ and Compact RIO chassis. It has 32 single-terminal inputs or 16 analog differentials, 16-bit resolution, and a maximum sample rate of 250 kS/s. Each channel has programmable input ranges of ±200 mV, V ± 1, ±5 V, and ±10 V. To protect against signal transients, the NI 9205 includes up to 60 V of overvoltage protection between the input and common channels (COM). In addition, the NI 9205 features a dual channel barrier isolation plant for safety, noise immunity, and high voltage range in common mode. It is rated for 1000 Vrms protection against transient overvoltages. The parameters acquired and linked to the NI 9205 module for analysis and monitoring are the AC, DC current and voltage signals, temperature signals, and radiation.
- The NI 9225 Series C analog input module has a total measuring range of 300 Vrms for high voltage measurement applications such as power metering, power quality monitoring, motor testing, battery testing, and batteries made out of fuel. The parameter acquired and linked to the NI 9225 module for analysis and monitoring, is the DC voltage signal that comes directly from the PV array.

8.3 Virtual Instrumentation (VIS)—Electrical and Environmental Parameters Analysis

Figure 8.2 shows the VIs front panel developed in LabVIEW [1] for the AC voltage and current signals, acquired by the real-time monitoring system. This VI has the option of selecting other sub-VIs such as power, phasor diagrams, and power quality, among others, in order to follow and monitor the system implemented in real time.

The purpose of this VI, "phasor analysis" as shown in Fig. 8.3, is to show the vector system's magnitude, angle, and impedance of the current and voltage.

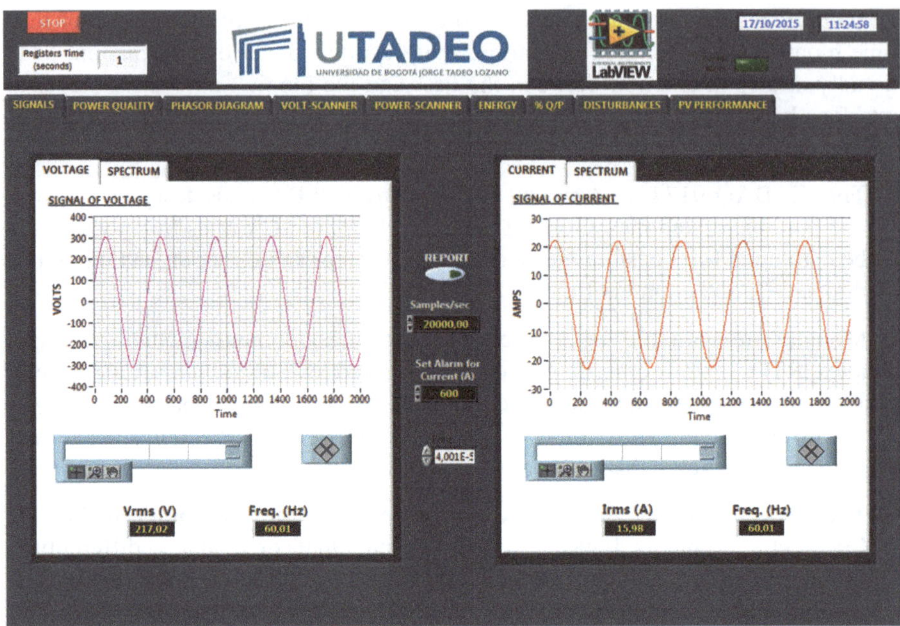

Fig. 8.2 VIs acquisition and monitoring of voltage and AC current

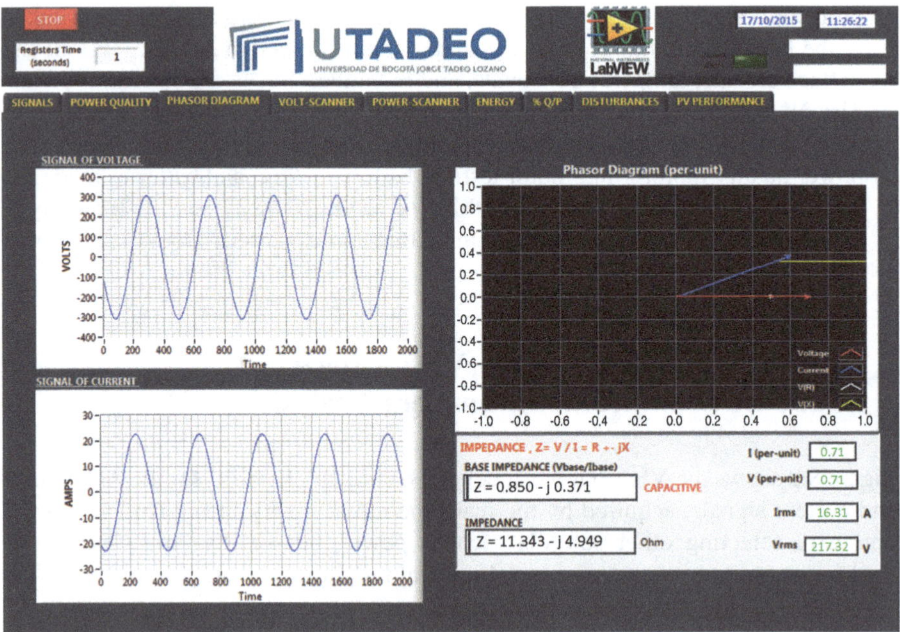

Fig. 8.3 VIs acquisition and monitoring of the phasor analysis

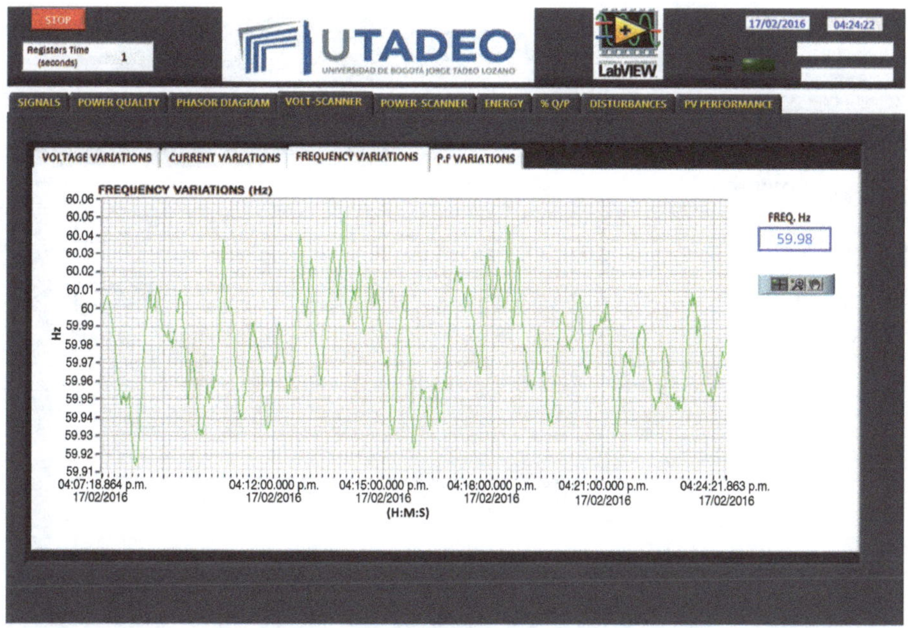

Fig. 8.4 VIs acquisition and monitoring of frequency behavior

In the Fig. 8.4, the behavior of the electrical signals is monitored and analyzed as the variation of the frequency, it is observed that the nominal frequency is within the limits required for its optimum functioning operation (59.8 and 60.2 Hz), as it is required by the standard.

The power factor of the system analyzed, as presented in Fig. 8.5, is at 0.75°, violating the requirements of the standard that it should operate above 0.85° in leading or lagging, for such a reason the system must be compensated with a bank of capacitors to improve its power factor and avoid penalties by the grid operator.

In Fig. 8.6 can be seen different types of electrical power of the system like the apparent power (S), reactive power (Q), and active power (W), since they are powers representing the description of the electrical system signals. These types of power are affected by the harmonic content caused by nonlinear loads or direct current to AC conversion equipment.

The active energy (Wh), Fig. 8.7, is directly linked to the active power (W) and the unit of time (t) in performing a job; The active energy is the description of the energy consumption required by an electrical system to supply the required demand.

Figure 8.8 shows the power quality parameters like total percentage of harmonic voltage distortion, voltage efficacy value, power factor, frequency, harmonic components, total harmonic current distortion percentage, current efficacy value, Active, apparent, reactive and distorting power, form factor, crest factor, and K factor, among others. This VI was developed for control and monitoring according

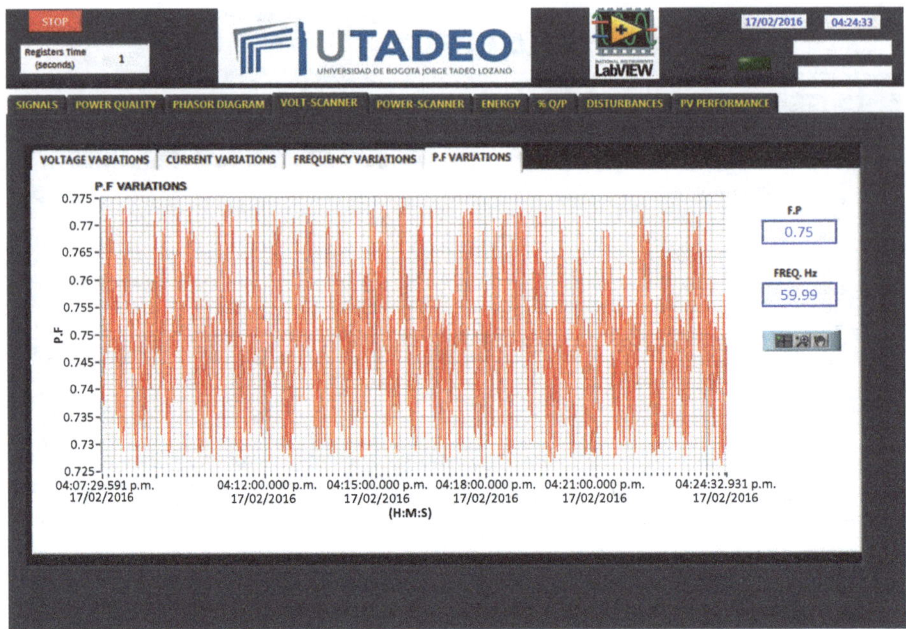

Fig. 8.5 VIs acquisition and monitoring of the power factor

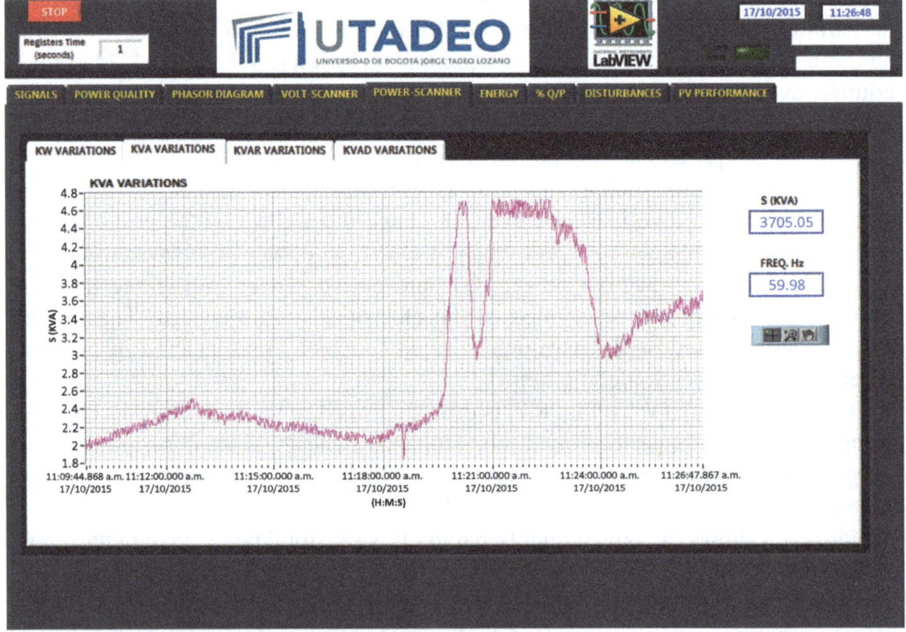

Fig. 8.6 VIs acquisition and monitoring of electrical power

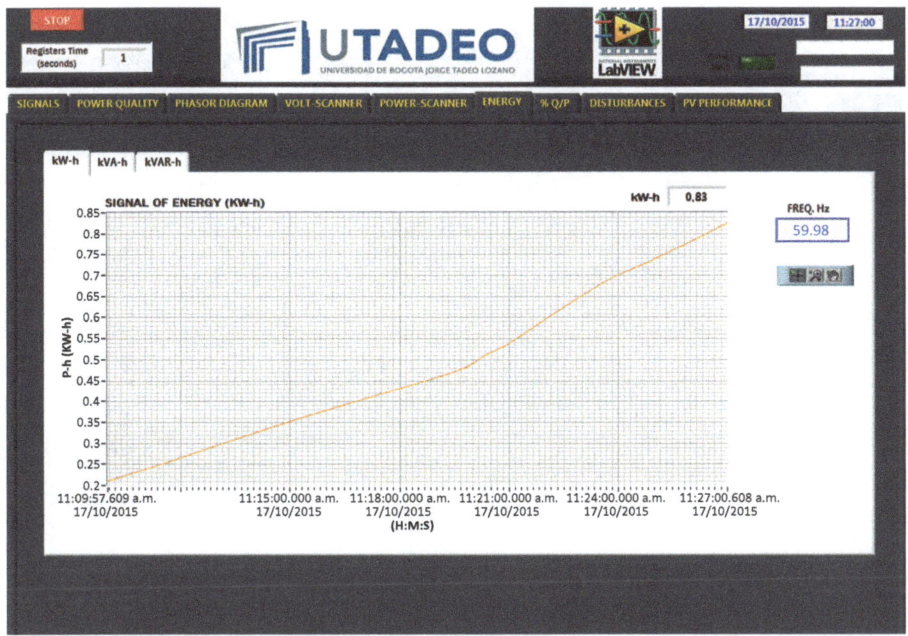

Fig. 8.7 VIs acquisition and monitoring of active energy

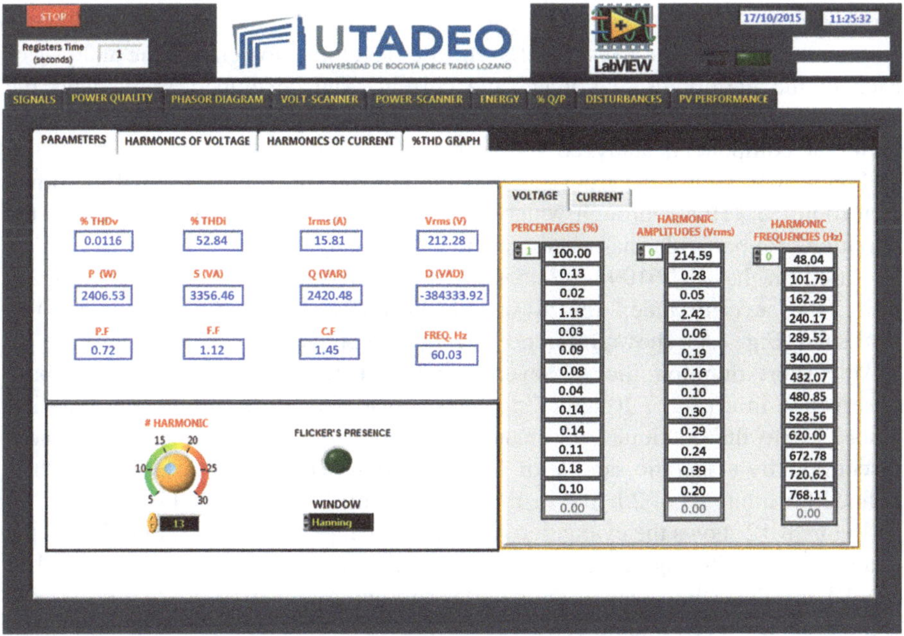

Fig. 8.8 VIs acquisition and monitoring of power quality

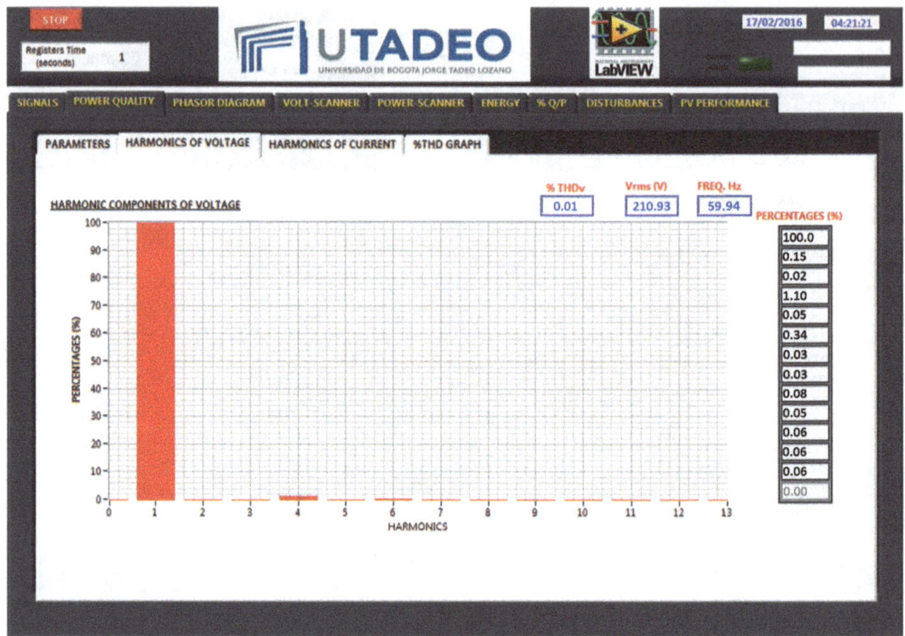

Fig. 8.9 VIs acquisition and monitoring of the voltage harmonic component

to the IEEE 929-2000 standard (power quality parameters) of the electrical system implemented at UJTL.

Figure 8.9 shows the front panel of the VI, developed to perform the analysis of each of the harmonics of voltage and current. The developed VI displays the spectrum or bar graph, which shows the influence percentage of each of the harmonic components analyzed.

Figure 8.10 shows the front panel of the percentage of voltage total harmonic distortion (% THDv) through which it is possible to analyze the behavior of the variation of a particular harmonic, for both the voltage signal and the current.

Based on the IEC 61000-4-7-15 standard, related to electromagnetic compatibility, the severity index for flicker of short (Pst) and long duration (Plt) is evaluated, Fig. 8.11 shows reference levels for voltage fluctuations; those are set by the short duration flicker Severity Index (Pst), which is defined for base observation intervals of 10 min. Pst = 1 is considered the threshold of irritability, associated to the maximum luminance fluctuation that can be supported without discomfort by a specific sample of the population. The long-term indicator Plt is defined for intervals of 2 h and its maximum permissible value is Plt = 0.74.

Figure 8.12 shows the evaluation of the technical performance of the BIPVS, the evaluation, monitoring and control of the parameters shown in the front panel of this VI, are the efficiencies of the inverter and photovoltaic generator and AC and DC currents, voltage and power.

The development of this VI, Fig. 8.13, allows for monitoring and evaluating the global solar radiation and the ambient temperature of the BIPVS installed at UJTL

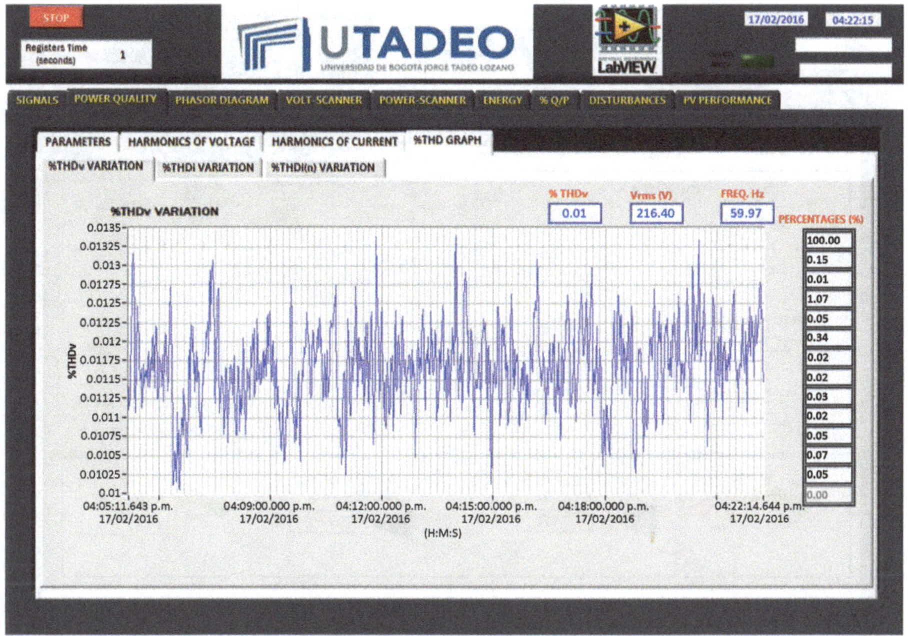

Fig. 8.10 VIs acquisition and monitoring for THD %

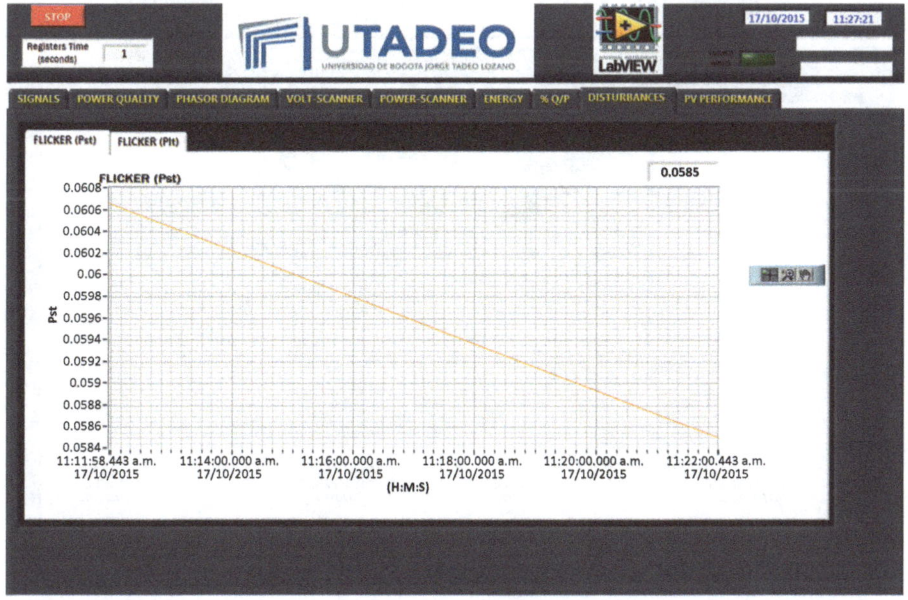

Fig. 8.11 VIs severity index for short duration flicker (Pst)

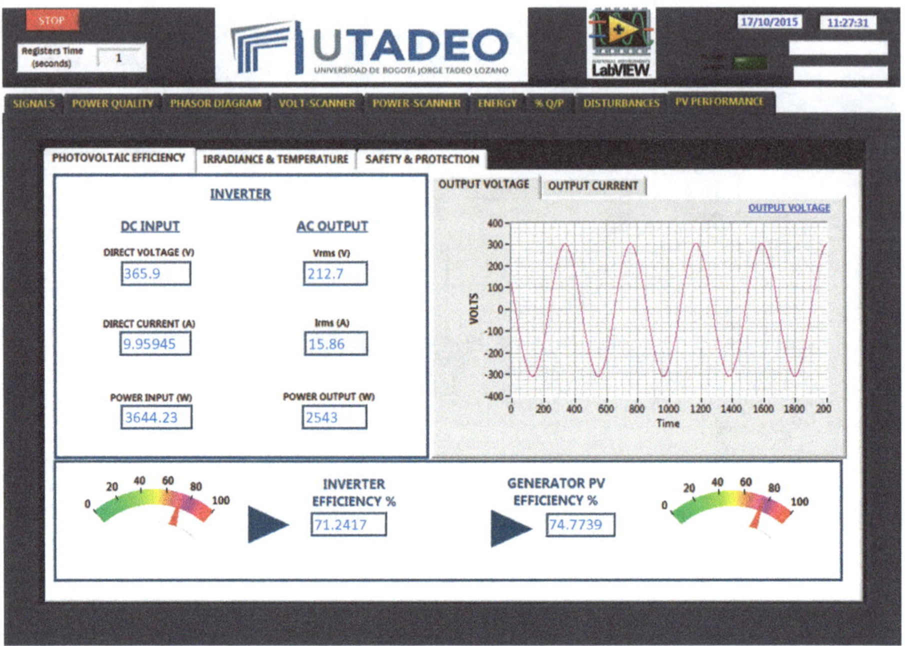

Fig. 8.12 VIs front panel for BIPVS performance

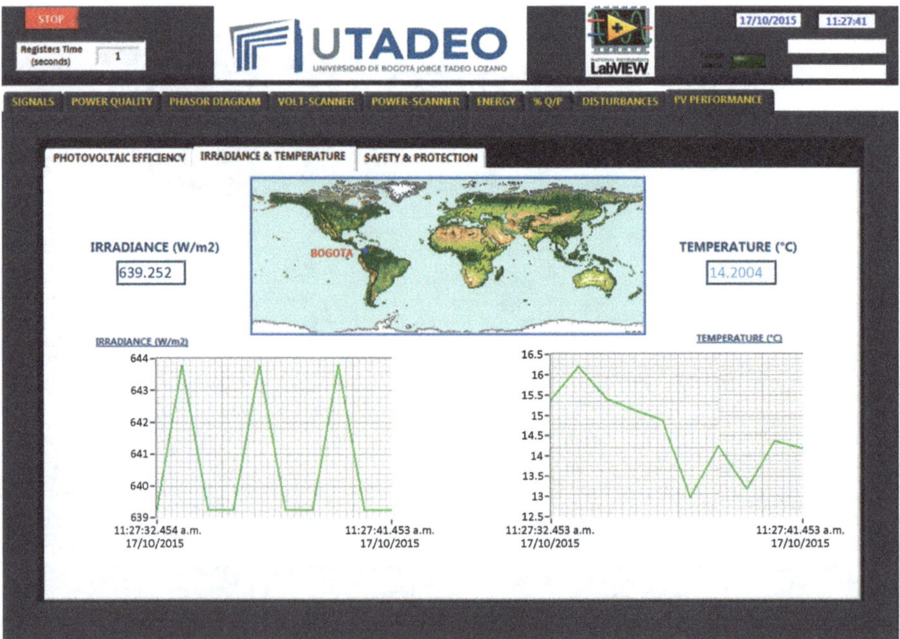

Fig. 8.13 VIs front panel for BIPVS environmental performance

in the city of Bogota. These parameters are necessary to calculate the conversion efficiency of the generator and the optimal sizing of a photovoltaic system.

In this VI can be observed the instantaneous numerical values and their real-time graphic representation of ambient parameters acquired and monitored like the solar radiation and ambient temperature of the implemented system.

Reference

1. G. Johnson, R. Jennings, *LabVIEW Graphical Programming*, 3rd edn. (Ed. McGraw-Hill, Austin, 2006), pp. 7–9

Chapter 9
Performance, Behavior, and Analysis of BIPVS

The BIPVS is installed in the upper platform of the CIPI (Center for Research and Engineering Processes) at the Universidad de Bogota Jorge Tadeo Lozano. It started operations and monitoring process since January, 2015 to date.

The performance, behavior, and analysis of the BIPVS will be evaluated from the technical aspect (BIPVS electrical and power quality parameters), the energy aspect (energy efficiency of the PV generator and the complete BIPVS) and from the environmental aspect (Environmental parameters like temperature and solar radiation).

These parameters are described in the IEC 61724 standard [1], these have been adopted by the international scientific community and are generally reported in monthly and/or annual periods.

9.1 BIPVS Technical Performance

9.1.1 PV Generator Performance

The main results and most relevant aspects of PV generator performance monitoring are presented in Figs. 7.1 and 7.2.

The total DC energy production (Fig. 9.1) of the PV array was 16.194 kWh during 2 years of operation; with a mean daily energy of 674.77 kWh. February, April and August were the months with the highest DC output energy.

Figure 9.2 shows that the PV array efficiency recorded an average of 13.75% for the 2 years of monitoring; with a minimum of 12.56% in the month of March, 2015; and a maximum of 14.87% for the month of April 2015. These results are due to the influence of temperature on the solar modules and changing conditions of cloudiness on the city of Bogotá.

February, 2015, June, 2015, April, 2016 and October, 2016 are in general the months with the highest solar irradiance in Bogotá with values of around 4500 Wh/m

A.J. Aristizábal Cardona et al., *Building-Integrated Photovoltaic Systems (BIPVS)*,
https://doi.org/10.1007/978-3-319-71931-3_9

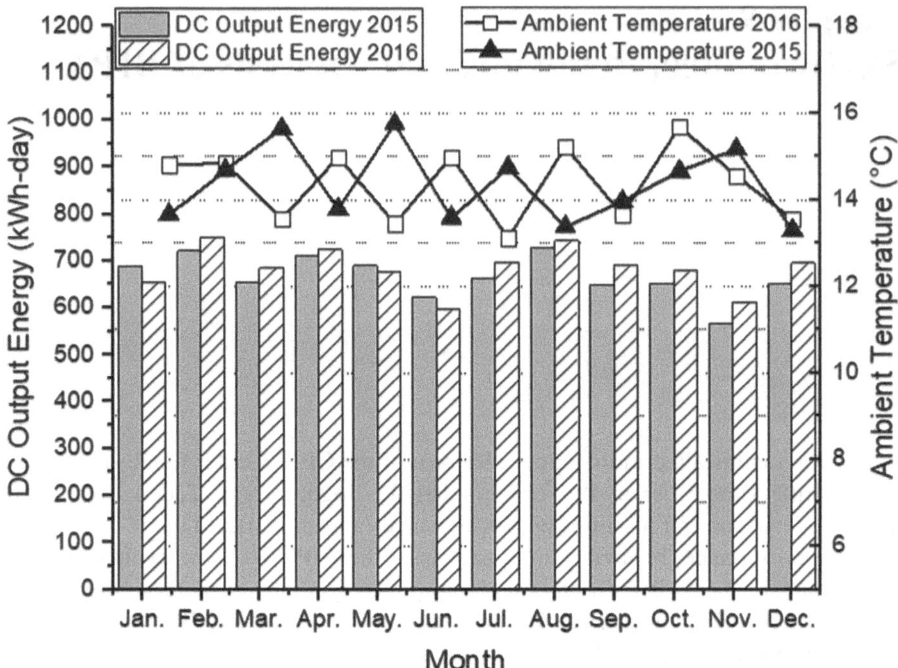

Fig. 9.1 Profile of the mean daily DC energy production of the PV array together with the mean daily ambient temperature, in Bogotá from 2015 (January) to 2016 (December)

^{2}-day; whereas March, July, 2015, August and December 2016 are the months with the lowest average of daily irradiance, with values around 3500 Wh/m^{2}-day. The annual daily average of solar irradiance in Bogotá was 4.06 kWh/m^{2}-day which is a low value for a region near the equator. This particular behavior seems to arise from the fact that Bogotá is located in the Andean Region, characterized by sudden atmospheric changes and intermittent cloud cover.

9.1.2 Inverter Performance

The power production generated by the inverter is shown in Fig. 9.3 along with the solar radiation profile for a sunny day between 6:00 am and 6:00 pm.

There is a high radiation (1300 W/m^{2}) with changes that could be construed as the emergence of small clouds during short periods of time due to the climate behavior in Bogotá city.

This clearly shows a similar pattern between the AC power generated by the inverter and solar radiation; obviously due to the PV generator's photocurrent dependence on the incident photons energy. This bell-shaped behavior indicates a radiation increase between 10:00 am and 3:00 pm, and then it begins to fall to zero at the 5:00 pm approximately.

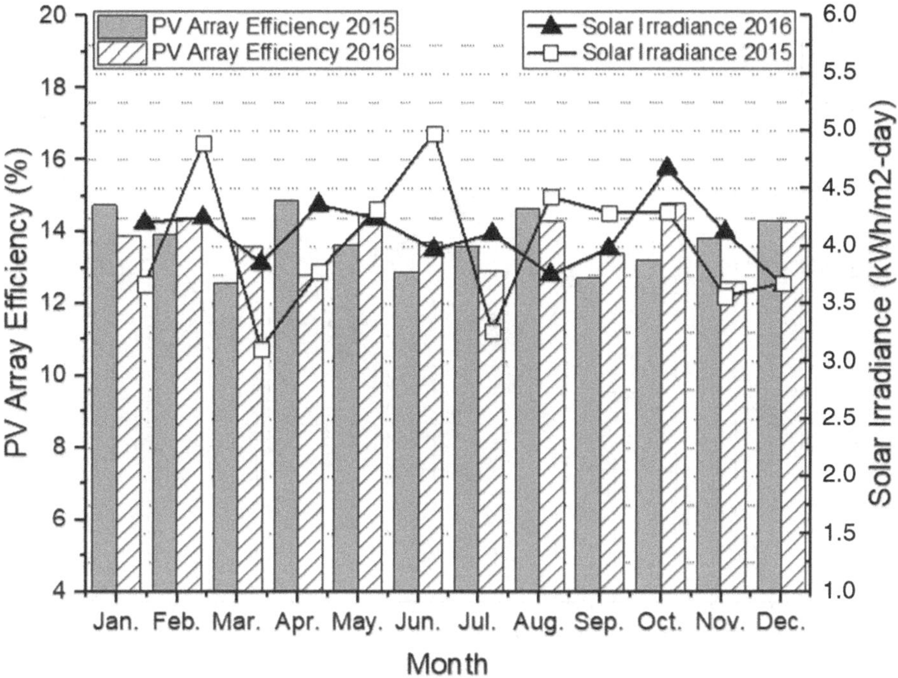

Fig. 9.2 Profile of the PV array conversion efficiency together along with the mean daily solar radiation in Bogotá, for 2015 (January) and 2016 (December)

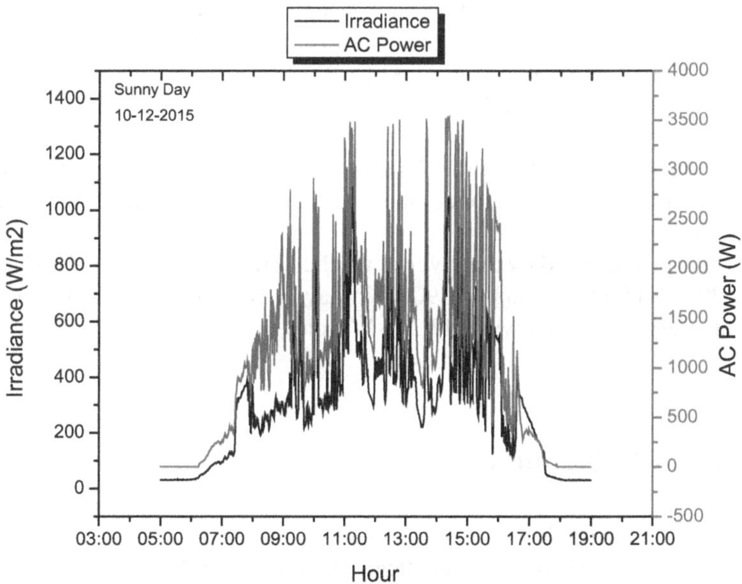

Fig. 9.3 System's irradiance and AC power variation for a sunny day

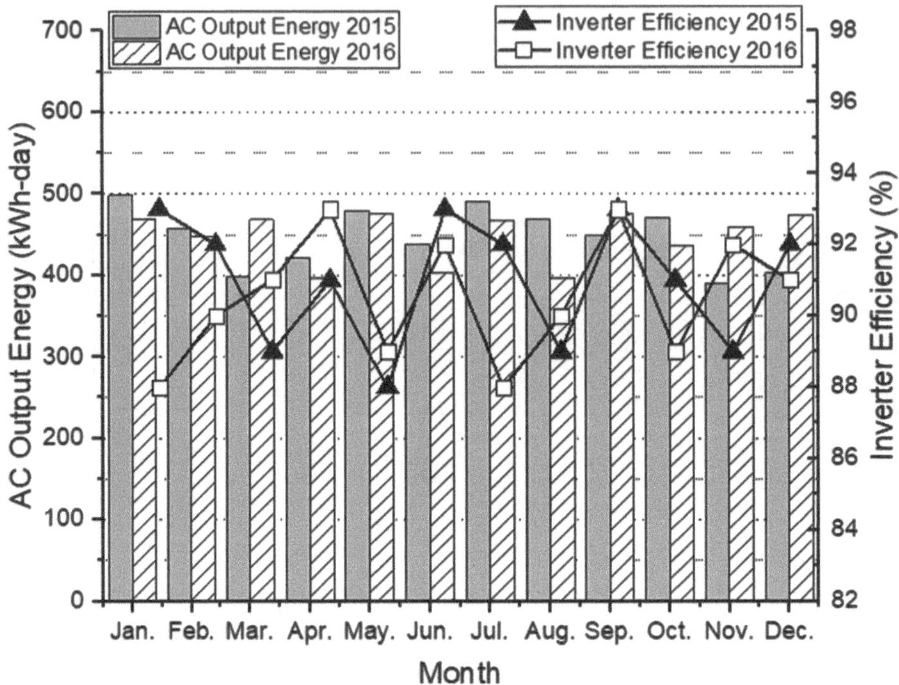

Fig. 9.4 SB 5000TL-US Inverter Efficiency and AC output energy

It is also clear that power is not generated in the early morning hours due to the fact that solar panels produce a voltage lower than 183 V which is not enough for the inverter to turn on and start the power delivery. Around 7:30 am, the inverter starts its operation, delivering power close to 600 W.

Figure 9.4 compares the monthly inverter's efficiency average values, calculated from the data obtained with the AC output energy. The results reveal a monthly average efficiency of 90.75% for the 2 years of monitoring varying between 88% and 93%, which indicates a photovoltaic system's correct operation achieving the maximum possible performance.

The AC energy produced by the inverter reaches a maximum of 498.56 kWh-day in January 2015 and a minimum of 390 kWh-day in November 2015. The average of ac output energy recorded was 447.24 kWh-day.

9.1.3 BIPVS Performance

Table 9.1 shows the BIPVS energy performance in accordance with the European recommendations and the IEC 61724 standard, and YA, YR, YF, PR, LC and LS parameters for 2-years of monitoring period.

Table 9.1 Values of: total energy produced by the BIPV system, parameters YR, YA, YF, PR, and LC, LS losses

Month	AC output energy (kWh-day)	Final yield YF (h)	PV array capture losses LC (h)	System losses LS (h)	Performance ratio (PR)	Reference yield YR (h)	Array yield YA (h)
January 2015	498.56	78.5	14.6	40.5	0.57	123.5	98.6
February 2015	456.89	90.3	23.5	36.4	0.68	112.6	105.4
March 2015	398.56	68.5	12.8	29.7	0.64	138.7	94.6
April 2015	421.45	76.4	17.8	36.8	0.58	111.2	117.4
May 2015	479.41	84.7	24.4	28.1	0.66	119.6	103.7
June 2015	437.8	68.2	12.7	38.9	0.63	125.8	98.3
July 2015	489.75	77.5	19.7	27.1	0.56	109.3	107.4
August 2015	469.41	73.2	22.3	39.6	0.6	116.7	103.2
September 2015	449.58	74.4	23.8	30.6	0.58	128.81	105.03
October 2015	470.65	79.7	24.5	29.2	0.6	133.41	108.8
November 2015	390	67.7	10.5	28.7	0.63	106.93	96.4
December 2015	402.35	68.9	12.3	41.6	0.56	122.91	110.6
January 2016	468.96	79.9	16.1	34.5	0.61	130.41	114.4
February 2016	448.21	71.2	14.5	37.1	0.58	123.12	108.3
March 2016	468.36	82.3	16.2	35.6	0.67	134.2	116.7
April 2016	395.45	84.6	16.7	34.8	0.63	129.4	118.4
May 2016	476.38	88.55	15.8	33.2	0.58	132.5	120.2
June 2016	402.54	68.5	11.3	29.7	0.67	116.7	92.8
July 2016	468.25	67.8	12.7	39.6	0.54	118.3	96.7
August 2016	396.25	73.2	22.5	31.2	0.63	122.6	105.8
September 2016	475.45	89.6	24.6	40.2	0.59	108.6	113.2
October 2016	435.69	63.2	16.8	37.6	0.68	122.8	118.1
November 2016	459.35	78.5	19.3	24.8	0.64	117.8	96.5
December 2016	474.36	84.3	11.3	29.2	0.57	127.5	104.3
Total	*10,733.66*	*1839.65*	*416.7*	*814.7*	*14.68*	*2933.39*	*2554.83*

The total annual final yield YF for the PV system installed was 1839.65 h. This value is according to those ones obtained for grid connected PV systems installed in several IEA-PVPS countries, which varies between 700 and 1840 h.

The reference energy performance YR recorded a value of 2933.39 h. The lowest value is recorded in the month of November of 2015 with 106.93 h and the highest in the month of March of 2015 with 138.7 h. The performance of the PV generator YA, recorded 2554.83 h and the high and low registers of the same performance are of 120.2 and 92.8 h corresponding to the months of May and June 2016. The calculated performance ratio of the PV system varies between 0.54 and 0.68 along the 24 months that we have monitored it.

During this period of 2 years, it is observed that the DC energy supplied by the photovoltaic system was 16.194 kWh equivalent to 16.2 MWh. The AC power of the system had a contribution of 10,733.66 kWh, lower than the DC power, due to the capture and system losses.

Figure 9.5 shows the PV array efficiency, inverter efficiency and system efficiency for the 24 months of monitoring.

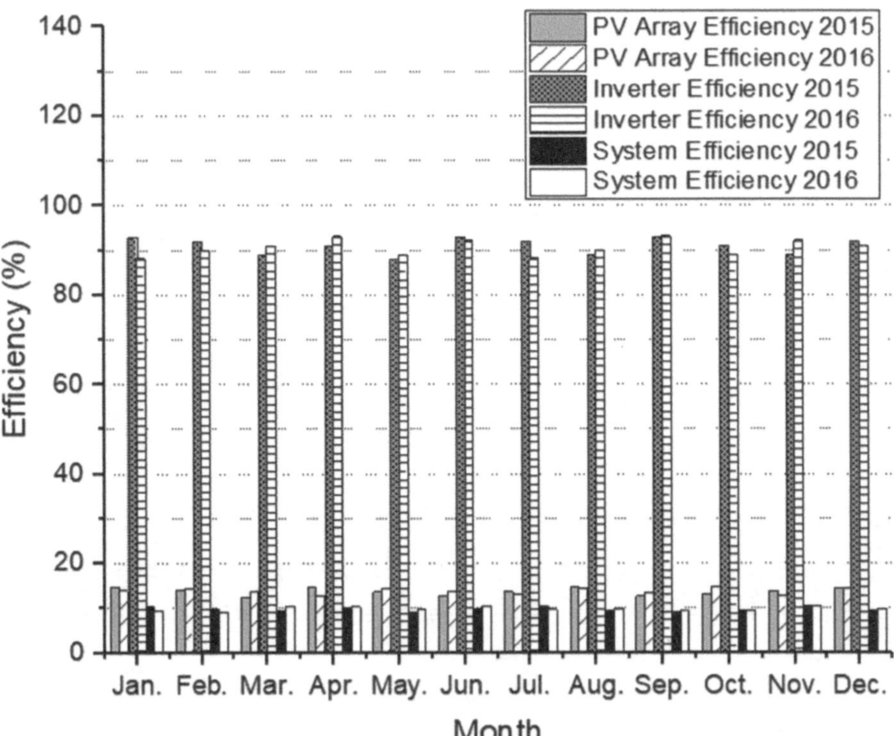

Fig. 9.5 Variations of: PV array efficiency, inverter efficiency and PV system efficiency

The efficiency of the inverter is above 90%. The efficiency of the PV generator is close to 15% and the efficiency of the complete system is close to 11%. This last efficiency is affected by the losses related to the inverter transformation of DC to AC power, as well as the dissipation of power in the cables carrying the energy between the PV generator and the electrical connection board.

The parameters that characterize the energy performance (PV generator performance YA, final yield YF, Reference yield YR and performance ratio PR) are defined in the document "European Guidelines for the Assessment of Photovoltaic Plants" [1]. These parameters are also determined from the measurements made with the monitoring system.

9.1.3.1 Photovoltaic Generator Performance ("Array Yield") YA

It is defined as the ratio of the energy produced by the photovoltaic generator in a given period of time (EGFV,τ) with respect to the installed nominal power.

$$\mathrm{YA} = \frac{\mathrm{EGFV}, t}{\mathrm{Pnom}, G} \left({}^{\mathrm{KWh}}, {}^{\mathrm{DC}}\!/_{\mathrm{KW,DC}}\right) \tag{9.1}$$

9.1.3.2 Final Yield YF

It is defined as the ratio between the useful energy produced by the PV system over a certain period of time, (EFV,τ AC), with respect to the nominal power installed.

$$\mathrm{YF} = \frac{\mathrm{EFV}, t, \mathrm{AC}}{\mathrm{Pnom}, G} \left({}^{\mathrm{KWh}}, {}^{\mathrm{AC}}\!/_{\mathrm{KW,DC}}\right) \tag{9.2}$$

9.1.3.3 Reference Yield YR

It is defined as the quotient between the solar radiation incident on the generator over a certain period of time (Ginc,τ) and the standard solar radiation GSTC (GSTC = 1000 W m^{-2}).

$$\mathrm{YR} = \frac{\mathrm{Ginc}, t}{\mathrm{GSTC}} \left({}^{\mathrm{KWh}}, {}^{\mathrm{m}^2}\!/_{1\ \mathrm{KWm}^2}\right) \tag{9.3}$$

9.1.3.4 Performance Ratio PR

It is defined as the ratio between the final yield, YF and reference yield, YR. This parameter is independent of the size (power) of the installation and is affected by

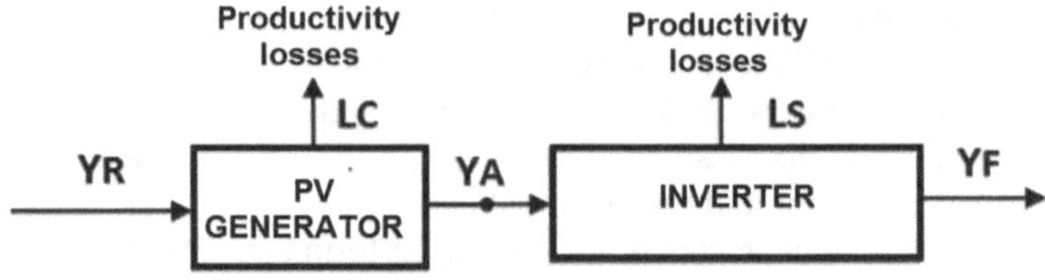

Fig. 9.6 Performance of the PV generator and rest of the system

the evolution of the ambient temperature; Allows to compare the behavior of different systems, in regards to the use of available solar resource.

$$PR = \frac{YF}{YR} \tag{9.4}$$

The block diagram in Fig. 9.6 shows the performance of energy productivity and losses associated with the entire generation system.

9.1.3.5 Photovoltaic Generator's Capture Productivity Losses (Lc)

These are due exclusively to the generator of the system and are a consequence of temperature in the cells, voltage drops in the wiring, dirt, shading, etc. [2].

$$Lc = YR - YA \tag{9.5}$$

9.1.3.6 Rest of the System Losses of Productivity (Ls)

Caused by subsystem inefficiencies (inverter and transformer), losses in AC wiring, maintenance shutdowns, faults or problems in the grid, among others [2].

$$Ls = YA - YF \tag{9.6}$$

Figure 9.7 shows the final yield performance (YF) of the BIPV system and the losses Lc and Ls during the 2 years of monitoring.

Figure 9.7 shows that the final yield performance YF of the BIPV system varies between 63.2 and 90.3 kWh/kWp. The LC losses of the PV array, recorded 416.7 h. LS losses, recorded 814.7 h.

Figure 9.8 shows the Performance Ratio (PR) of the generation system from January 2015 to December 2016.

The calculated performance ratio (PR) of the PV system varies between 0.54 and 0.68 along the 24 months that have been monitored (Table 9.1). The low PR values obtained can be attributed to the power losses caused by the PV system.

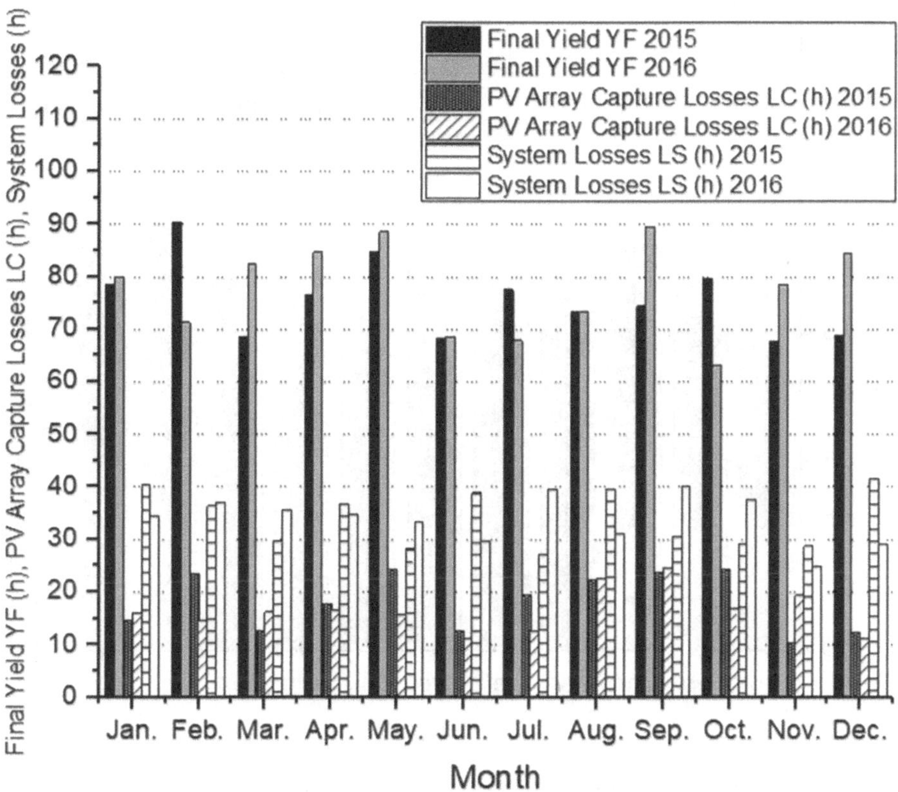

Fig. 9.7 Performance parameters: YF, Lc, Ls of the BIPV system

Figure 9.9 shows data between January, 2015 and December 2016 and Table 9.1 show data for 2 years where the reference energy performance YR recorded a value of 2933.39 kWh/kWp. The lowest value is recorded in the month of November 2015 with 106.93 kWh/kWp-year and the highest in the month of March 201 with 138.7 kWh/kWp-year. The performance of the PV generator YA, recorded 2554.83 kWh/kWp and the highest and lowest registers of the same performance are 120.2 and 92.8 kWh/kWp-year corresponding to the months of May 2016 and June 2016.

The total annual final yield for the PV plant installed at the Universidad de Bogota Jorge Tadeo Lozano was 1839.65 kWh/kWp. This value is according to those ones obtained for grid connected PV systems installed in several IEA-PVPS countries, which varies between 700 and 1840 kWh/kWp-year.

It can be seen that the yield YF is lower than the other yields YR and YA due to losses occurring in the system as seen in the block diagram of Fig. 9.6.

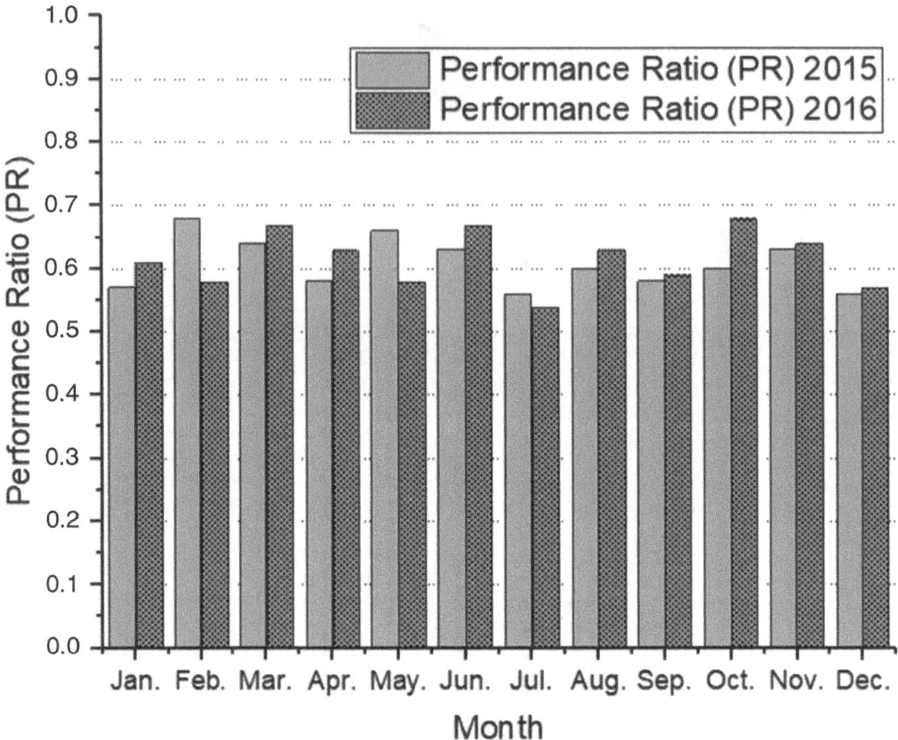

Fig. 9.8 BIPVS PR performance ratio

9.2 BIPVS Performance and Analysis of the Power Quality

To know the behavior and use of the power quality in the BIPVS, it is necessary to use measurements and analysis of the electrical parameters stored by the acquisition system. The basic parameters of electrical energy such as voltage and current can undergo changes, generating disturbances in the electrical systems, affecting the quality of the power energy service.

Figure 9.10 shows the frequency and voltage level of the BIPV system for the 2 years of monitoring.

Figure 9.10 shows that the frequency was always within the permitted limits with a higher frequency of 60.03 Hz and a lower frequency of 59.974 Hz. These results meet IEEE Standard 929-2000.

In the same Fig. 9.10, the lowest effective voltage of 212.54 V was recorded in the month of May 2015, while the highest value recorded was in the month of April 2015 with 216.98 V. It is observed that the stability voltage always remained within the limits set by the standard: +5% and −10%.

In Fig. 9.11 the behavior of the flickers and the power factor for the same time period is shown.

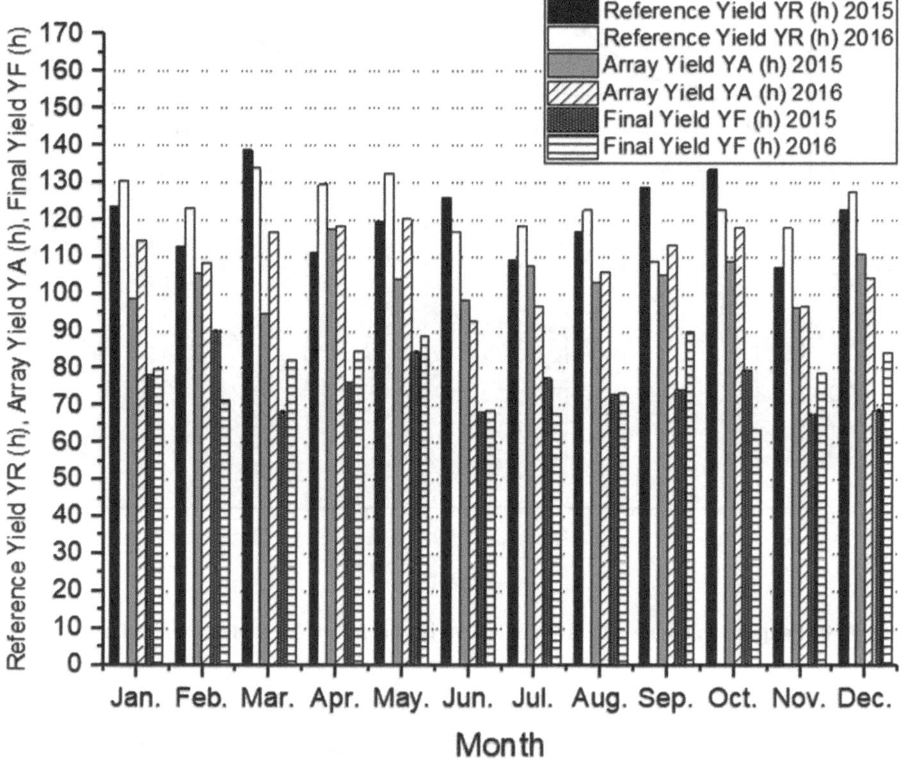

Fig. 9.9 Energy Performance YR, YA, YF of BIPVS

Figure 9.11 shows the variations of Flicker (short duration) (Pst) did not reach the limit established by standard (Pst = 1). The lowest record was 0.0630 captured in the month of October 2016 and the highest of 0.0890 in August 2016.

As for the power factor, it shows that it is not complying with the limits of the IEEE 929-2000 standard, as this record must be greater than 0.85, and the recorded data was always under. This result is due to the presence of capacitive and inductive loads in the building; for the most part by the installation of laboratory equipment. To improve and correct the power factor must be set and design a capacitor set to a power factor of 0.92, which must be connected to the main bus bar of the PCC.

Figure 9.12 displays the variation of the AC energy generated and the behavior of the percentage of the voltage total harmonic distortion (% THDv) for the months of January 2015 through December 2016.

Figure 9.12 shows the variation of voltage total harmonic distortion (% THDv). According to the standard, it should always be less than 5%. The value of this parameter always remained below the 2.12% whose registration was presented in July 2015 and the lowest value was registered with a value of 1.02% in the month of January 2015.

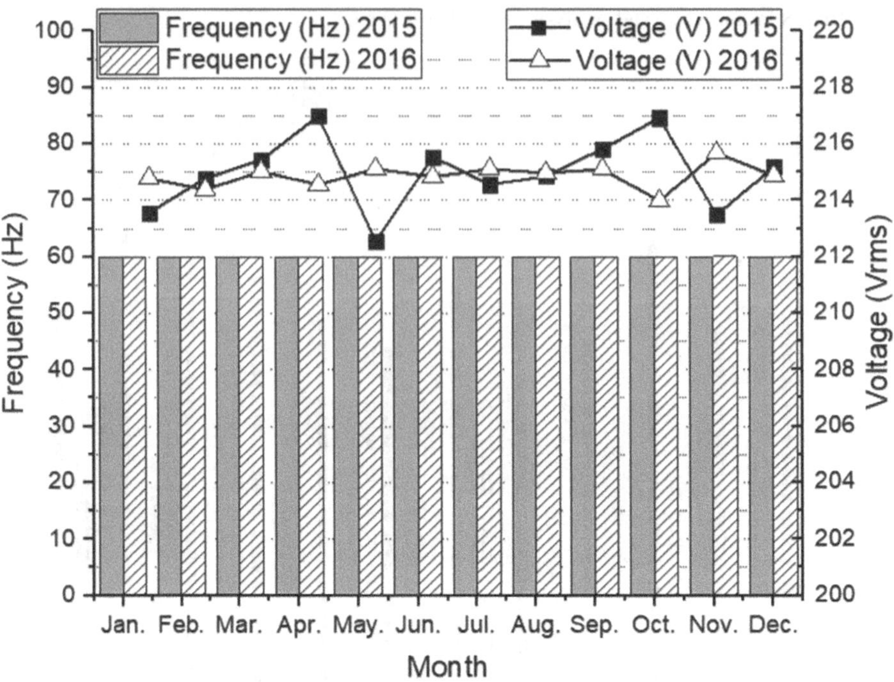

Fig. 9.10 Frequency and the effective voltage variations for the 2 years of monitoring

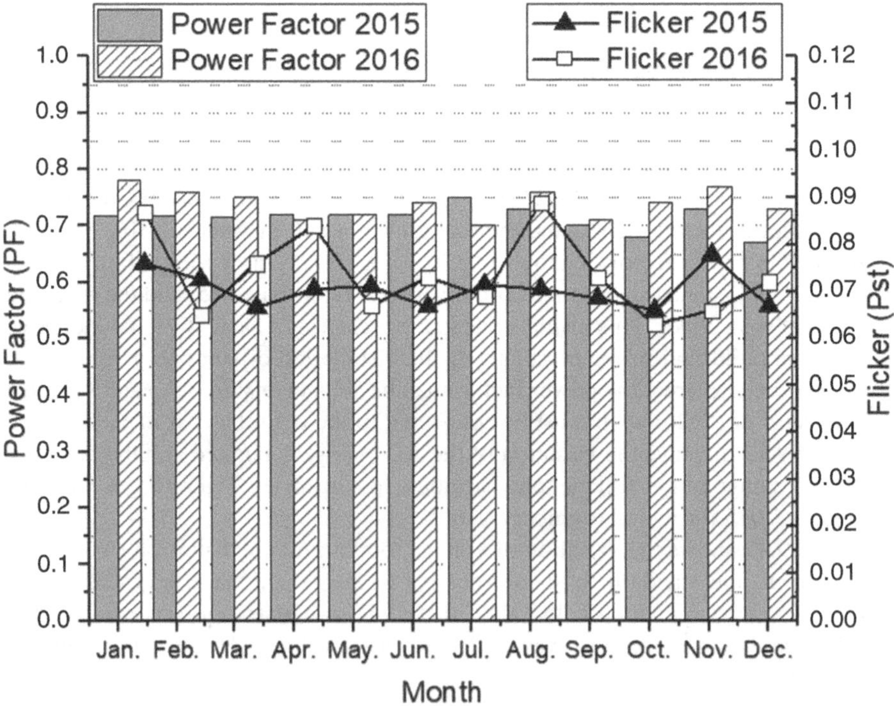

Fig. 9.11 Power factor and short-term flicker (Pst) variations for 2 years of monitoring

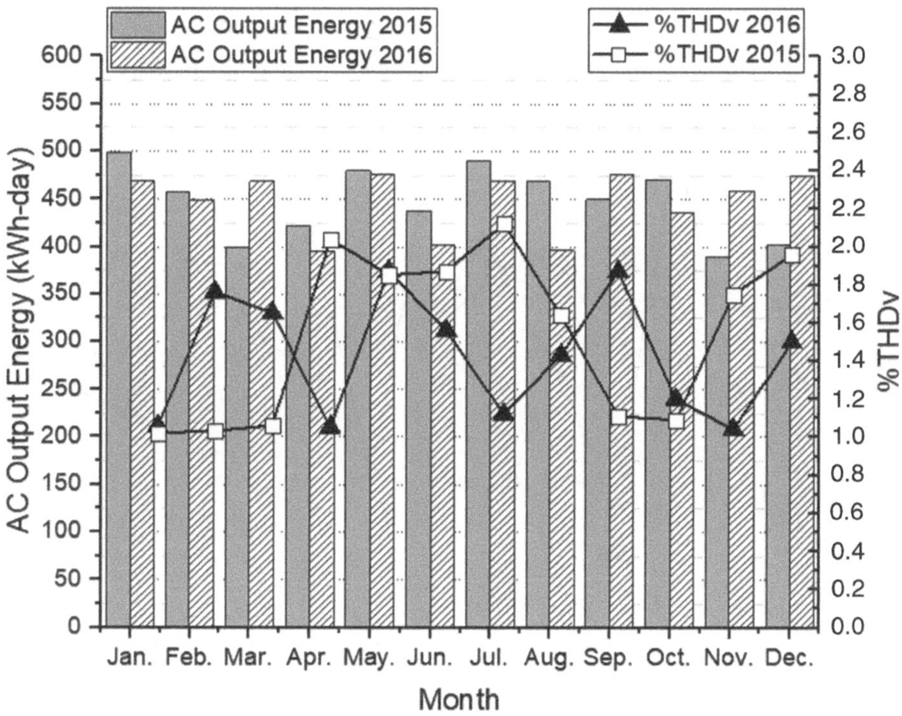

Fig. 9.12 Variation of generated AC energy and behavior of voltage total harmonic distortion

9.3 BIPVS Environmental Performance

Below is the data acquired by the monitoring system such solar radiation, ambient temperature, and the average hourly radiation and temperature values measured since January 2015.

9.3.1 Solar Radiation Data

It is basically used to obtain information required for the sizing of photovoltaic systems. The data analyzed and monitored are the average daily and monthly values of global solar radiation and the hours of standard sunshine. Figs. 7.13 and 7.14 summarize the results of the monitoring process.

It is observed in Fig. 9.13 that the months of greatest solar radiation were October 2015, February 2016 and June 2016 which reached values higher than 380 W/m^2 and the ones with the lowest solar radiation were March, July November and December 2016 in which the average daily radiation was just above 250 W/m^2.

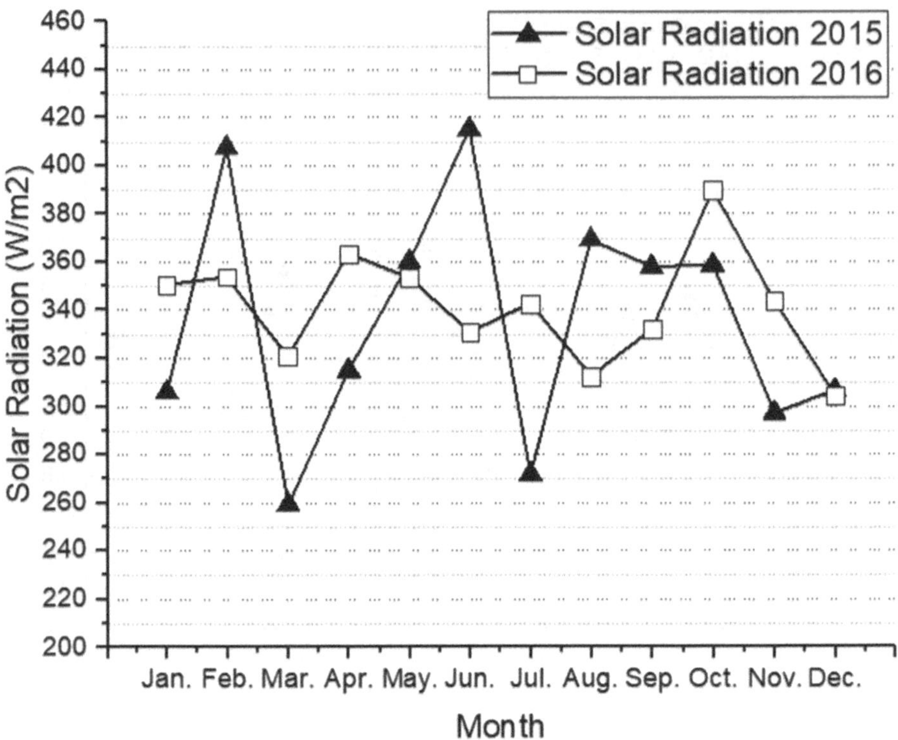

Fig. 9.13 Average daily monthly irradiance in the city of Bogota

These results show a cyclical behavior of global solar radiation as in other research literature [2].

Figure 9.14 shows the graph of the daily profile of solar irradiance corresponding to December 28, 2015, a day that was characterized by a high level of radiation. In the graph, the numerical values of the average solar radiation and the maximum radiation recorded are labeled. The average irradiance recorded was 219 W/m^2 and the maximum of this day was 1171 W/m^2, with changes that could be interpreted as the emergence of small clouds during short periods of time due to the climate behavior in Bogota city.

9.3.2 Ambient Temperature Data

The data analyzed and monitored are the average daily and monthly values of ambient temperature recorded at the Universidad de Bogota Jorge Tadeo Lozano. Figures 9.15 and 9.16 summarize the results of the ambient temperature monitoring process.

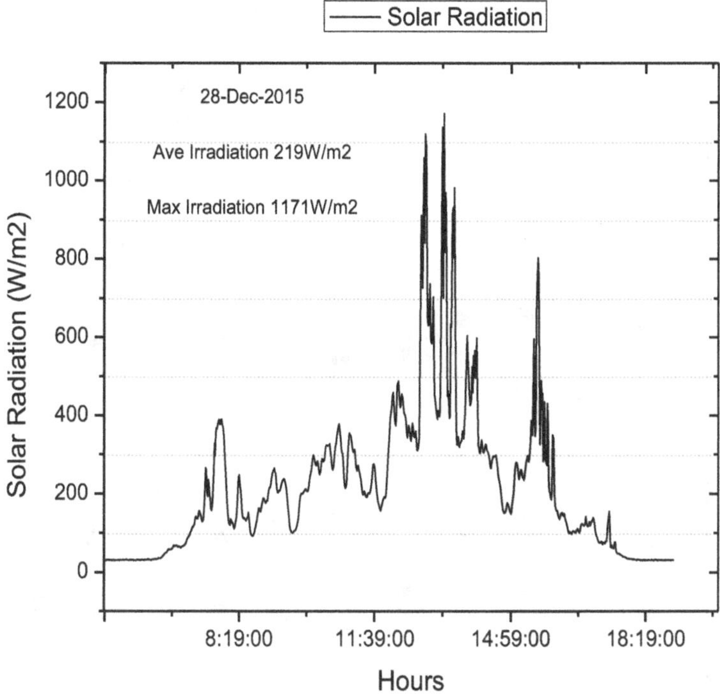

Fig. 9.14 Daily profile of solar irradiance. December 28th, 2015

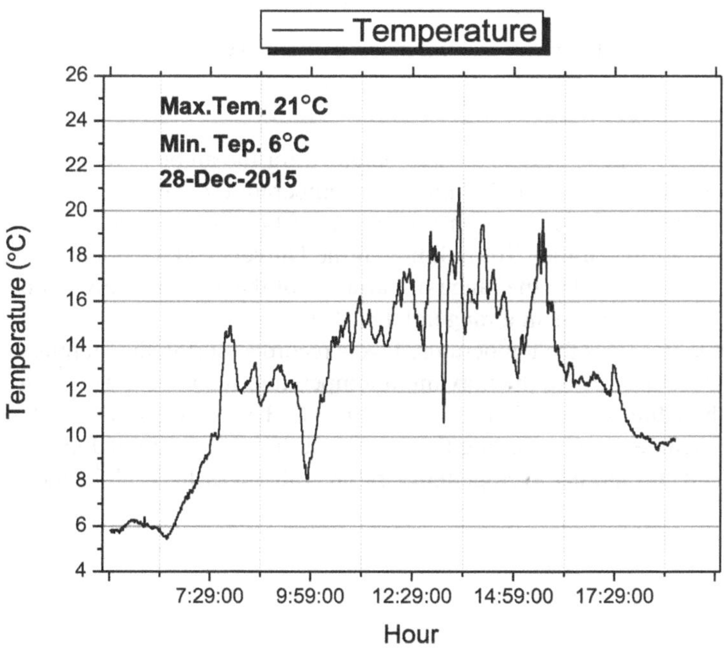

Fig. 9.15 Daily profile of ambient temperature. December 28th, 2015

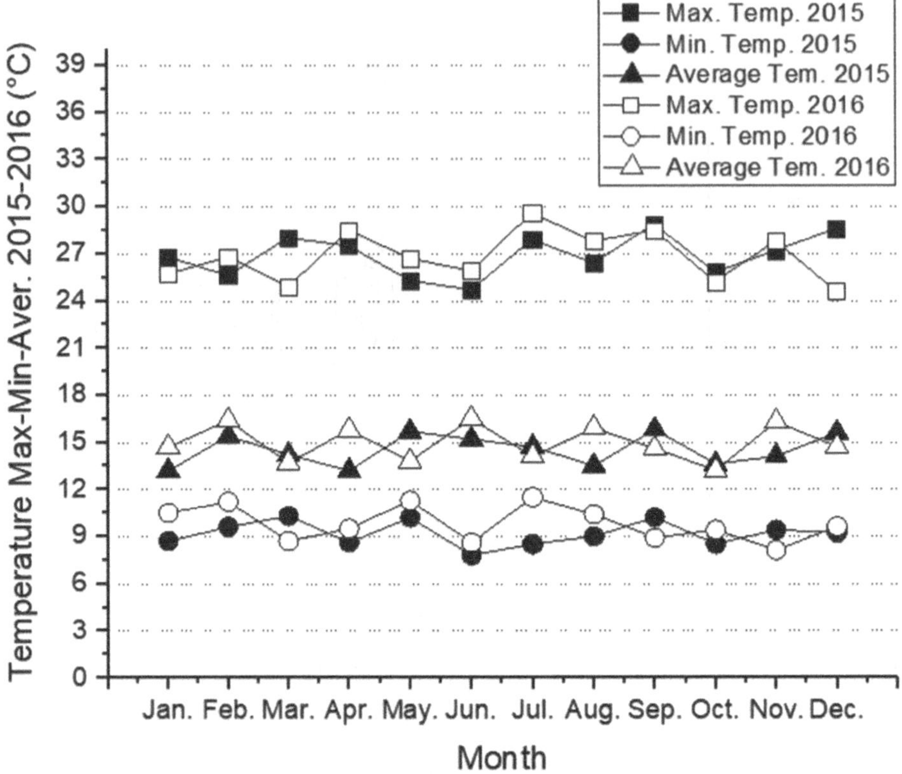

Fig. 9.16 Average daily monthly ambient temperature at UJTL in the city of Bogota 2015–2016

In Fig. 9.15 a general register was obtained for the ambient temperature with a maximum value of 21 °C and a minimum temperature record of 6 °C. This record refers to December 28, 2015, with a time zone of 14 h of monitoring and location on the upper platform of the CIPI building at the Universidad de Bogota Jorge Tadeo Lozano. This means that the range of variation of the average daily temperature of the sector for this day is in a range of 12.38 °C.

Figure 9.16 shows the temperature levels recorded during the months of January 2015 to December 2016, the maximum temperature recorded was obtained during the month of July 2016 with a record of 29.6 °C. The same analysis is done for the minimum temperature recording that occurs during the month of June 2015 with a value of 7.8 °C, and the average temperature value was 14.7 °C for the 2 years.

References

1. IEC Std 61724, *Photovoltaic System Performance Monitoring Guidelines for Measurements, Data Exchange, and Analysis*
2. E. Kymakis, S. Kalykakis, T. M. Papazoglou, *Performance Analysis of a Grid Connected Photovoltaic Park on the Island of Crete* (2009)

Chapter 10
Behavior and Analysis of the Power System in Steady State

The Electrical System needs to remain at all times in a stable equilibrium state, where generation covers demand instantaneously, and where its most important variables do not exceed certain limit values. The system operator is responsible for maintaining the correct functioning of the system through the coordination between generation, transport and distribution so that the electricity supply is made in a safe, reliable and with the required quality [1].

The purpose of this chapter is to determine by means of simulations the analysis and behavior of the power system operation in steady state, incorporating a photovoltaic generator in the main distribution board of the CIPI building. To support this purpose it is used an ETAP software, version 12.6.0, which is a tool for analysis, design, simulation and operation of electrical power systems. The developed simulations are: power flow analysis, short circuit calculations and analysis of harmonics.

The stability of a power system is defined as the ability of the system, for a given initial condition, to recover a normal state of equilibrium after being subjected to a disturbance [2, 3].

- Single-line diagram of the power system

The single-line diagram shown in Fig. 10.1 corresponds to the actual electrical system implemented at the CIPI building at the Universidad de Bogota Jorge Tadeo Lozano in the city of Bogota. The electrical power system consists of a connection point or external grid at a voltage level of 11,400 V, supplied by a pedestal type transformer 75KVA 11,400/208 V, with exclusive use at the UJTL. A low voltage cable with a length of 60 m, (3 $\times$ 4/0 F + 1 $\times$ 4/0 N + 1 $\times$ 2 T THHN), a photovoltaic generator of 6.2 KW, an inverter of 5 KW, the inverter output arrives at a Point of common coupling bar (PCC), where also the transformer output is connected, Which in turn supports the entire system load.

A.J. Aristizábal Cardona et al., *Building-Integrated Photovoltaic Systems (BIPVS)*,
https://doi.org/10.1007/978-3-319-71931-3_10

10.1 Power Flow

The concept of load flow or power flow is associated with the prediction of the flows direction in the grid, the voltages and powers, which allow to evaluate the state of the grid in advance and predict an inadequate situation in the system. In general, the power flow is a basic tool to determine the steady-state operating conditions of a power system [4].

The load flow is performed for the planning, design and operation of power systems under various operating conditions and to study the effects of changes in configurations in the equipment, in addition to contingency analysis, optimum dispatch, and stability, among other studies [5, 6].

A load flow calculation determines the state of the power system for a given load and power distribution. This represents a steady state condition as if that condition had to be kept fixed for some time. The load flow problem can be defined as the calculation of the active and reactive powers flowing in each branch and the magnitude and phase angle of the voltages of each bar in a power system for given load and generation conditions [5, 6].

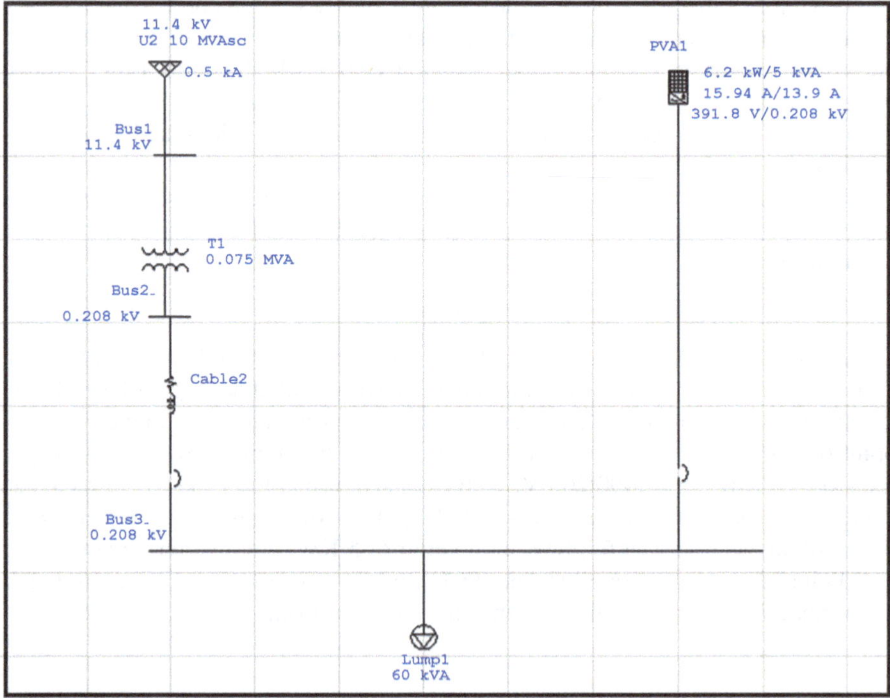

Fig. 10.1 Single-line diagram of the power system at CIPI Building—UJTL

10.1.1 Load Flows Analysis

Figure 10.2 shows the simulated electrical system under the load flow parameter, the data supplied by the program indicates: that the level of voltage at medium voltage present on bus 1 is kept constant at 11,400 V. Active and reactive power is found within limits required by the system (48 KW and 30 KVAr). Bus 2 has a voltage level of 203.5 V equivalent to a voltage drop of 2.16% of the initial reference voltage (208 V), active and reactive power At this point it has the following values (47 KW and 29 KVAr). Bus 3 has a voltage level of 199.1 V equivalent to a voltage drop of 4.27% of the initial reference voltage and the active and reactive power in Bus 3 corresponds to the sum of the power supply of the photovoltaic generator (5 KW and 0 KVAr) and the initial electrical system (47 KW and 29 KVAr) to obtain a power demanded by the system connected to the general distribution board or PCC, equivalent to (52 KW and 28 KVAr = 60 KVA).

This indicates that the system is operating within the requirements of voltage levels in the bus bars and active and reactive power generation in each branch of the system according to the requirements of the IEEE 929-2000 standard regarding the parameters of power quality that governs the national territory, like voltage stability since the effective value of the voltage must be kept as stable as possible (low

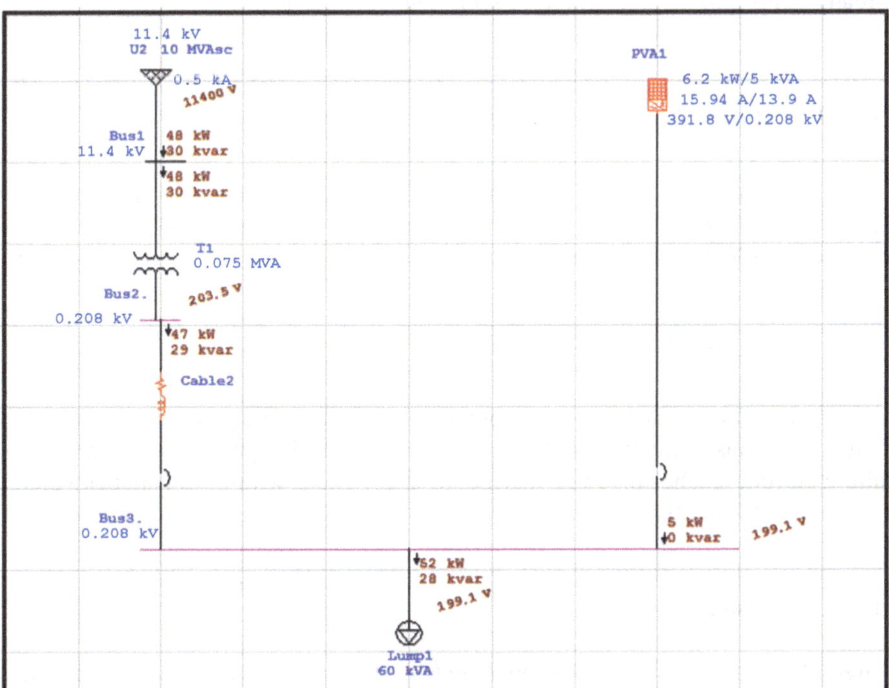

Fig. 10.2 Simulation of load flow with Etap

voltage +5% and −10%—medium voltage +5% and −10%.) And power factor greater than 0.85 in leading or lagging.

To improve and correct the voltage level in bar 3, it is advisable to change the position of the transformer tap to raise the voltage level and improve the stability on this bus.

10.2 Short Circuit Studies

A failure is defined as the passage of the electric current from an energized conductor, "phase," towards the ground or the neutral conductor. When this happens a study of the impedances of the elements that caused the fault is required [7].

Because the short circuits cause changes in the magnitudes of a power system, it is necessary to carry out a study of those to size and check protection devices (switches, circuit breakers), also determine the efforts and influence components of the system are subjected to under a short circuit in a specific area [8].

The studies about failure in a circuit have led to show great interest in the calculation of these, because they cause high currents, voltage drops, overheating and instability of the electrical system. The calculation of short-circuit failure is of great importance since they help the section of power switches and they must already withstand the magnitude of such high currents [9].

10.2.1 Short Circuit Analysis

Next, the short-circuit analysis is performed for different cases of three-phase failure that may occur in the electrical power system. Simulations are done under the ANSI C37/UL489 standard.

10.2.1.1 Case 1

Three-phase failure in bus 1: As shown in Fig. 10.3, the three-phase failure is simulated in bus 1, the voltage level becomes zero at this point, and the generated failure current reaches 0.515 KA, affecting bar 1 and 2 with instantaneous voltage drops of 0.019 and 0.033 kV.

10.2.1.2 Case 2

Three-phase failure in bus 2: As shown in Fig. 10.4, three-phase failure is simulated in bus 2, the voltage level becomes zero at this point, and the generated failure current reaches 5.3 KA, affecting bar 1 with instantaneous voltage of 9.514 kV

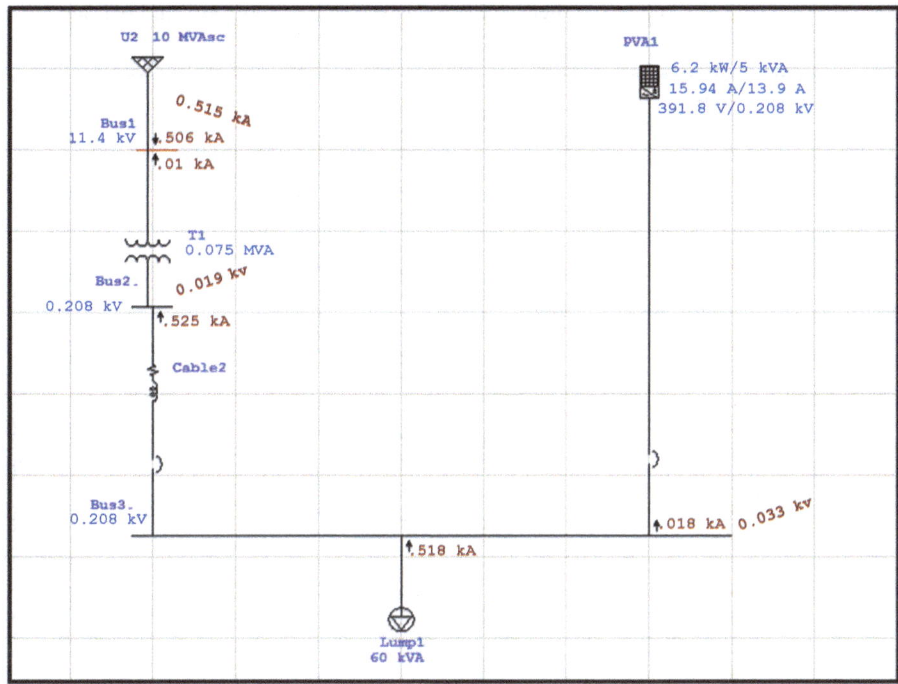

Fig. 10.3 Simulation of three-phase short circuit in bus 1

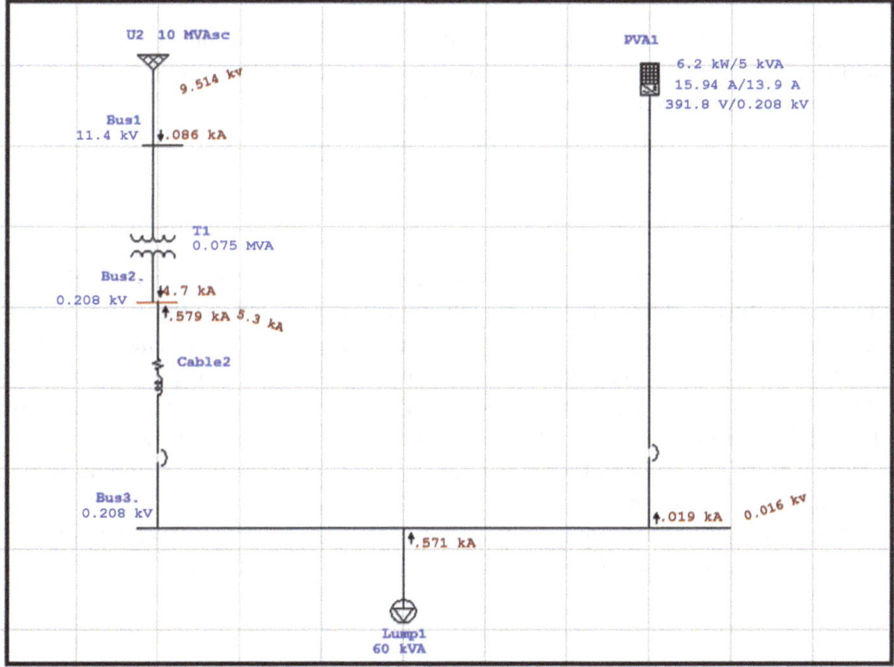

Fig. 10.4 Simulation of three-phase short circuit in bus 2

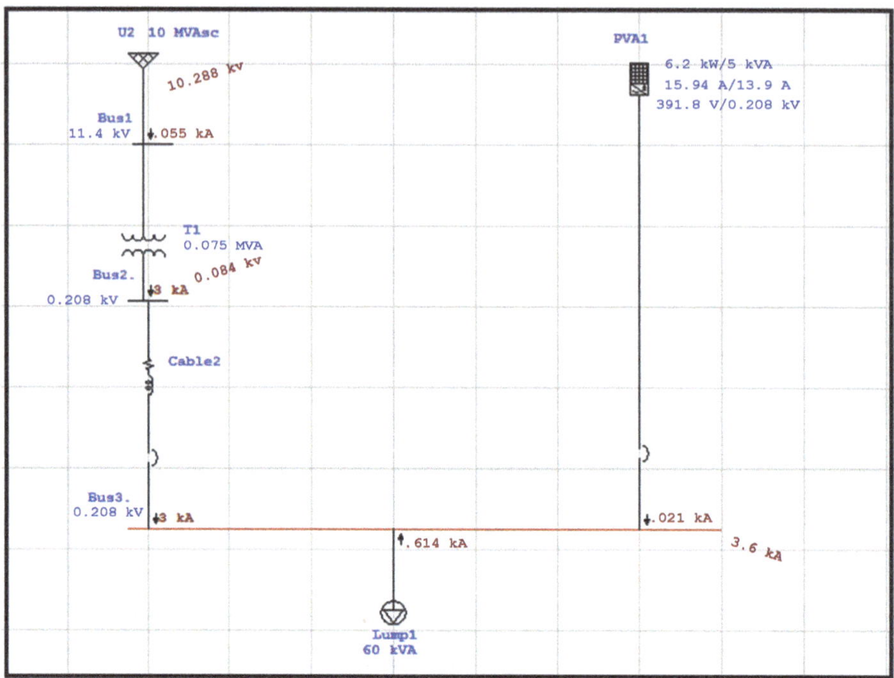

Fig. 10.5 Simulation of three-phase short circuit in bus 3

corresponding to 16.5% voltage drop and a current flow of 0.086 KA; The same effect occurs in bar 3 with an instantaneous voltage of 0.016 kV corresponding to 92.3% of voltage drop and a current flow of 0.552 KA.

10.2.1.3 Case 3

Three-phase failure in bus 3: As shown in Fig. 10.5, the three-phase failure is simulated in bus 3, doing the same analysis in this simulation, the voltage level becomes zero at this point (bus 3), and the generated failure current reaches 3.6 KA, affecting bar 2 with an instantaneous voltage of 0.084 kV corresponding to 60% voltage drop; The same effect occurs in bar 1 with an instantaneous voltage of 10.288 kV corresponding to the 9.75% voltage drop and a current flow of 0.55 KA.

With a more detailed and thorough analysis of 1/2, 4, 30 cycles, different single-phase and three-phase short-circuit current failures can be analyzed where design, selectivity, coordination, equipment acquisition and power system complexity requires it and where the interpretation depends on each particular system.

10.3 Harmonics Studies

The power quality can be quantified technically through the following parameters: Waveform, frequency, symmetry, reliability, voltage flicker, waveform distortion, and distortion power, among others.

The IEEE 519-1992 standard establishes that the total harmonic distortion must be less than 5% of the fundamental frequency in relation to the output of the inverter.

10.3.1 *Harmonic Analysis*

In the single-line diagram of Fig. 10.6, the case study of harmonic analysis is simulated by the Newton/Rapson method, with a maximum of 99 iterations, and a precision of 0.0001. The harmonic source model used is Thevenin/Norton equivalent for all Sources (method recommended by the software).

It is observed that the simulated data indicate that the instantaneous voltage levels are stable in bars 1, 2, and 3, 11.4–0.208—0.208 KV, respectively and that the total harmonic distortion (THD) at the same points present (1.89%—1.15%—

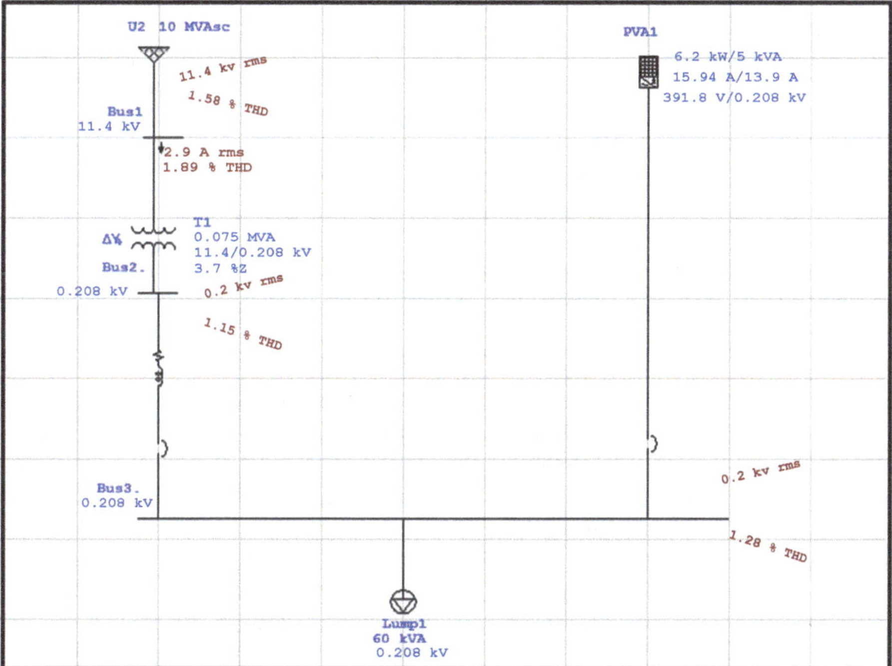

Fig. 10.6 Simulation of total harmonic distortion THD

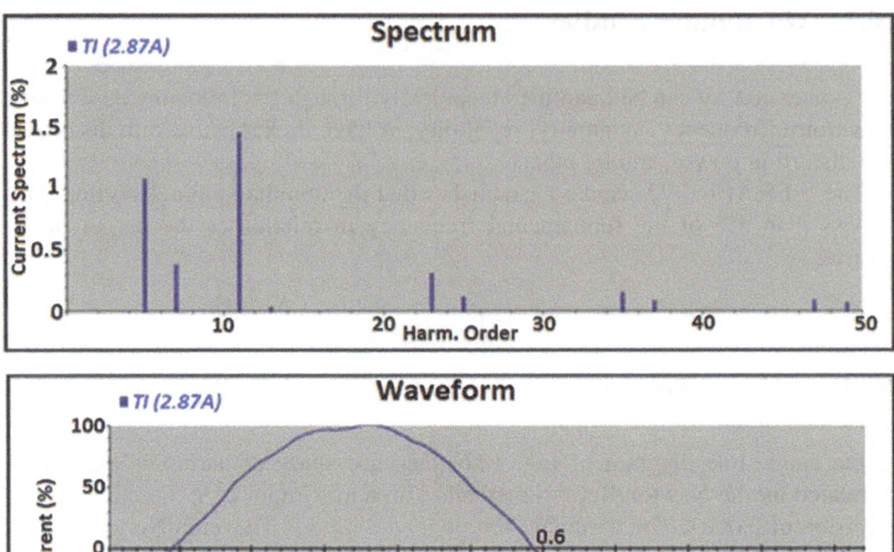

Fig. 10.7 Harmonic currents spectrum and transformer current wave

1.28%), indicating that parameters like voltage and THD are within the limits established by the aforementioned standard.

Case 1 THDi (%) Transformer: Fig. 10.7 presents the spectrum with the harmonic components of currents and the current wave that flows through the 75 KVA transformer. It is observed that the most appreciated harmonic current components are the odd components of order 5, 7, 11, 23 whose values are reflected in Table 10.1.

As shown in Table 10.1, the odd components of harmonic current do not exceed the distortion limits required by the IEEE 519-1992 standard.

Case 2 THDv (%) in Bars 2 and 3: Fig. 10.8 shows the spectrum of the voltage harmonic components and voltage waveform for bars 2 and 3. It is observed that the most appreciated voltage harmonic components in bar 2 are the odd components of order 11, 13, 23, 25, 35, and 37; in the same way, it is also analyzed that the harmonic components of bar 3 are almost identical to bar 2, the values of these components are reflected in Table 10.2. Also these odd voltage components do not exceed the limits of distortion required by the aforementioned standard.

Table 10.1 Odd component of harmonic current present in the 75 KVA transformer

S

Copy Format

#	X	Y1
0	1	0
1	3	0
2	5	1.07618
3	7	0.378238
4	9	0
5	11	1.45242
6	13	3.12448e-
7	15	0
8	23	0.312819
9	25	0.119864
10	35	0.159304
11	37	8.78309e-
12	47	9.34906e-
13	49	6.89945e-

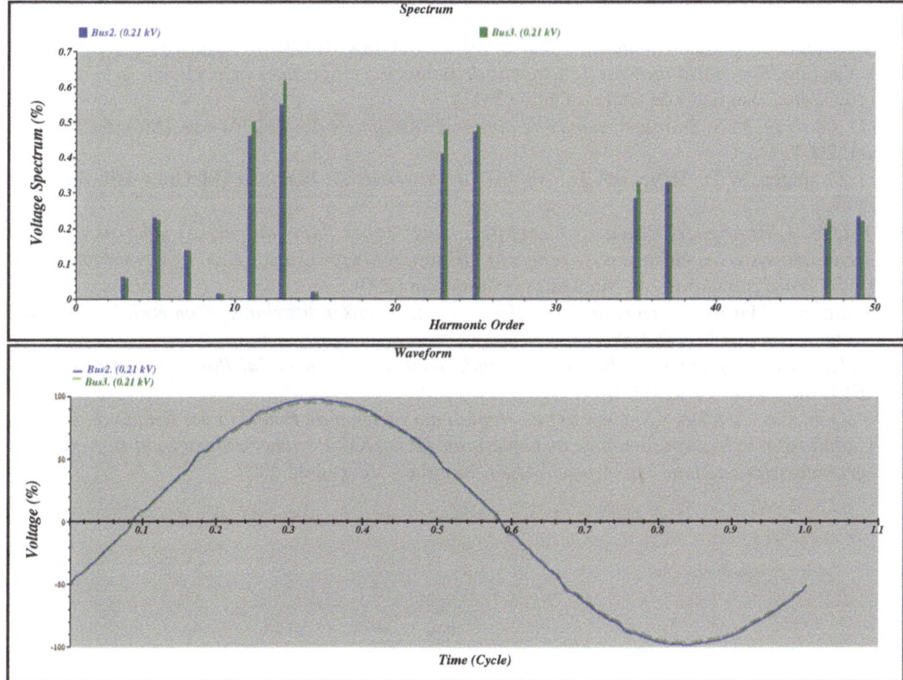

Fig. 10.8 Harmonic components voltage spectrum and voltagewave of bars 1 and 2

Table 10.2 Odd component of harmonic voltage in bars 2 and 3

#	X	Y1	Y2
0	1	0	0
1	3	6.52885e-	6.06968e-
2	5	0.230442	0.224389
3	7	0.142039	0.138061
4	9	1.85227e-	1.67663e-
5	11	0.463191	0.500768
6	13	0.550601	0.618316
7	15	2.24468e-	1.94647e-
8	23	0.411653	0.478728
9	25	0.474149	0.490877
10	35	0.284787	0.327876
11	37	0.331611	0.329713
12	47	0.20025	0.225558
13	49	0.232904	0.221308

References

1. R. Fuentes, *Herramientas de visualización de estudios de flujos de potencia para el apoyo a la toma de decisiones en sistemas eléctricos*. Santiago de Chile – Chile (2005)
2. P. Kundur, *Power System Stability and Control* (McGraw Hill, New York, 1994)
3. A. Castillo, *Respuesta inercial de sistemas de potencia con grandes inyecciones de generación fotovoltaica*. Santiago de Chile – Chile (2013)
4. J.D. Glovery, M.S. Sarma, *Sistemas de potencia: Análisis y diseño*, 3rd edn. (México, Thompson, 2004)
5. J.J. Grainger, W.D. Stevenson Jr., *Análisis de Sistemas de Potencia* (McGraw-Hill, México, 1996)
6. O. Días, J. Rodríguez, *Practicas Computacionales sobre flujos de carga y análisis de corto circuito en sistemas eléctricos de potencia utilizando los programas Etap Powerstation 3.02 y Power Word Simulator 7.0*. Manizales – Colombia (2001)
7. G. Enríquez Harper, *Protección de instalaciones eléctricas industriales y Comerciales*, 2nd edn. (Limusa, México D.F, 2003)
8. IEEE, *IEEE Recommended Practice for Industrial and Commercial Power Systems Analysis* (IEEE Inc., New York, 1990)
9. W. Gonzales, *Análisis y Simulación en Neplan del Sistema de Potencia del Estado de Barinas (Cadafe) con la Incorporación de un Generador en 115KV, Perteneciente al Central Azucarero Agroindustrial Ezequiel Zamora (CAAEZ)*. Merida – Venezuela 2007

Chapter 11
Application of Neural Networks to Validate the Power Generation of BIPVS

11.1 Development of a Neural Network to Evaluate the BIPVS

Since the BIPVS offers the possibility to replace part of the traditional building materials, with a possible price reduction, in comparison to a classic rooftop installation [1, 2], then the correct estimation of system level performances, system reliability and system availability is becoming more important and popular among installers, integrators, investors and owners; with that purpose several tools and models were developed [3–5]. The combination of different phenomena, such as solar radiation available on site, dust presence, shadowing or UV radiation over long outdoor exposure, affect in different ways the real performance of BIPV systems and thus the related economic evaluations [6–8].

A variety of circuit-based models, such as single diode model and double diode model, have been proposed to describe the *I–V* characteristics of a PV module [9–13]. The former techniques usually employ simplified formulas and information provided by the PV module manufacturer [14–16]. Hence, these techniques are easy to implement but more vulnerable to the accuracy of the few available data points. The latter techniques, including neural network [17, 18], differential evolution (DE) [19], genetic algorithm [20], and particle swarm optimization (PSO) [21], etc., usually employ all the experimental data at various operating conditions to extract physical parameters, thus provides a higher confidence level of the extracted parameters.

This chapter describes how to estimate the output power of the BIPVS with a mathematical model proposed through the implementation of an artificial neural network.

The grid-connected BIPV system installed at the Universidad de Bogota Jorge Tadeo Lozano and the monitoring system creates the necessary database for use in the proposed artificial neural network. Saved variables are: solar radiation, ambient temperature, DC current and voltage, and AC voltage and current. The monitoring

A.J. Aristizábal Cardona et al., *Building-Integrated Photovoltaic Systems (BIPVS)*,
https://doi.org/10.1007/978-3-319-71931-3_11

system saves 1 data per minute of each variable; between 5 am and 7 pm. Files are saved in .xls format on a daily basis. Solar zenith angle and solar azimuth angle are derived from time of day and day of the year by using the solar position algorithm developed by NREL. A 12-month (all 2016) monitoring data is used to train the model and do the performance testing.

11.1.1 Artificial Neural Network Model (ANN)

A typical feed forward neural network is depicted in Fig. 11.1:

This neural network can be used to fit any kind of finite input to output mapping problem with enough neurons in the hidden layer. The input layer accepts inputs of an M-dimensional vector, X; the hidden layer, H, is composed of N neurons; and the output layer, Y, has Q outputs. The connection between layers is as follows [22]:

$$h_j = f_H\left(\sum_{i=1}^{M} x_i \omega_{i,j} + b_j\right), \quad \text{for } j = 1, \ldots, N \tag{11.1}$$

$$y_q = f_Y\left(\sum_{J=1}^{N} h_j \omega'_{j,q} + b'_q\right), \quad \text{for } q = 1, \ldots, Q \tag{11.2}$$

where x_i, h_j, and y_q are, respectively, the ith input element, the jth output of the hidden layer, and the qth element of the output layer; $u_{i,j}$ is the weight from ith input element to the jth neuron of the hidden layer, and u_{0j},q is the weight from jth neuron

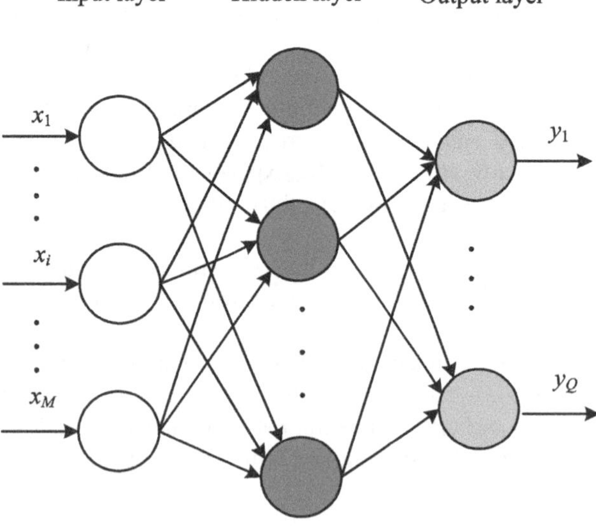

Fig. 11.1 Basic model of an artificial neural network [22]

of the hidden layer to the qth neuron of the output layer; b_j and b_{0q} are the biases of the jth neuron of the hidden layer and qth neuron of the output layer, respectively; f_H and f_Y are the transfer functions for the hidden layer and the output layer, respectively.

This model is used to estimate PV DC and AC energy from environmental information.

Before training the network, the input and output data sets are normalized to [1, 1] according to the following expression:

$$v = v_{\min} + \frac{u - u_{\min}}{u_{\max} - u_{\min}}(v_{\max} - v_{\min}) \tag{11.3}$$

Where u is the original data falling in the interval $[u_{\min}, u_{\max}]$, and v is the corresponding normalized value in the interval $[v_{\min}, v_{\max}]$.

The commonly used hyperbolic tangent sigmoid function and linear function are used as the transfer function of the hidden layer and output layer, respectively. The tangent sigmoid function is defined as [22]:

$$\text{tansig}(s) = \frac{2}{1 + \exp(-2s)} - 1 \tag{11.4}$$

where s represents the net input of a neural network node.

The Levenberg-Marquardt (LM) algorithm, a widely used back-propagation algorithm for supervised learning, is employed to update the network weights and biases in the following way:

$$\mathbf{w}_{k+1} = \mathbf{w}_k - \left(\mathbf{J}^{\mathrm{T}}\mathbf{J} + \mu\mathbf{I}\right)^{-1}\mathbf{E} \tag{11.5}$$

where $\mathbf{w}_k$. is the vector of weights and biases at step k, $\mathbf{J}$ is the Jacobian matrix of the network error $\mathbf{E}$ with respective to weights and biases, $\mathbf{J}^{\mathrm{T}}$ is the transpose of matrix $\mathbf{J}$, $\mathbf{I}$ is the identity matrix, and m is the damping parameter of the LM algorithm.

11.2 BIPVS Performance Results

To know the energy performance and quality of electrical power in the PV system it is necessary to make measurements and analysis of the electric parameters stored by the mentioned acquisition system. Basic parameters of power as well as voltage and current can suffer changes, creating perturbations in the electrical systems, affecting the quality service of electrical power. The AC output power, the PV array efficiency, and the solar radiation data for 2016 are taken from Chap. 9.

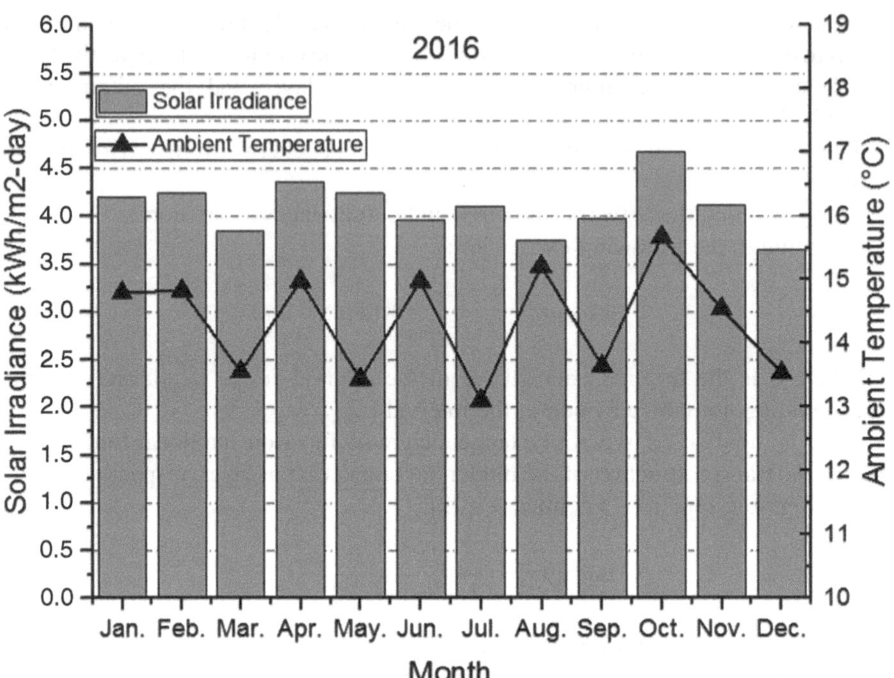

Fig. 11.2 Solar radiation daily average along with daily average monthly of ambient temperature for 2016

In Fig. 11.2, it is observed the solar radiation variation vs ambient temperature measured in Bogota.

April, May and October are the months with higher solar irradiance in Bogotá with values above of 4 kWh/m^2-day. The average irradiance was 4.09 kWh/m^2-day. The daily average monthly maximum temperature reached is presented in the month of June with around 16.5 °C, presenting the lowest average temperature in the month of October with 13.2 °C.

In Fig. 11.3, one can observe the daily average of solar AC output energy and the PV array efficiency measured in Bogota.

The ac energy produced by the PV inverter reaches a maximum of 476.38 kWh/month in May and a minimum of 395.45 kWh/month in April. The average of AC energy recorded was 475.23 kWh/month.

The PV array efficiency recorded an average of 9.86% for the 2016; with a minimum of 9.2% in the month of February and a maximum of 10.50% for the month of March. These results are due to the influence of temperature on the solar modules and changing conditions of cloudiness of the city of Bogota.

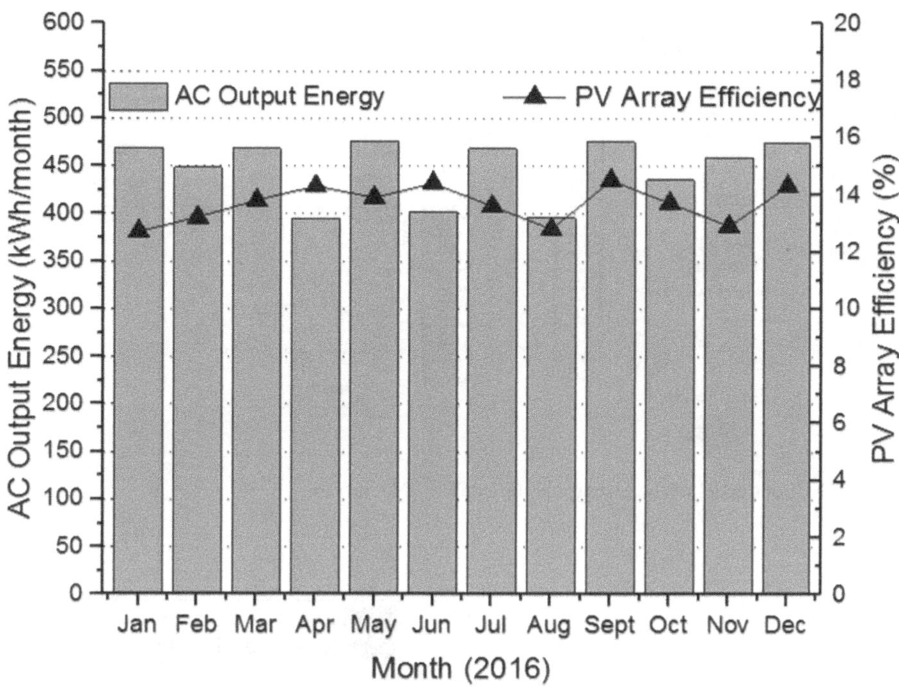

Fig. 11.3 Daily average of solar AC output energy along with PV array efficiency for 2016

11.3 ANN Model Results

The ANN (artificial neural network) model was developed in LabVIEW (see Fig. 11.4) so this program creates an .xls file with the results. The meteorological information to train the neural network corresponds to a year of data of solar radiation and ambient temperature acquired by the monitoring system during 2016. Using the described neural network, it is possible to model the behavior of the dc-ac power and energy of the BIPV system. The model's data is then validated with the data measured by the monitoring system.

After performing the respective tests, it was found that with 14 neurons it is enough to estimate the power and energy using four inputs: solar radiation, ambient temperature, azimuth angle and zenith angle. The neural network was trained using the group of measured data of the whole year 2016: one data per minute for 365 days (between 5 am and 7 pm; daily).

Figure 11.5 shows the results of the model applied to the ac power of the BIPV system for October 28, 2016.

The day analyzed was characterized by high cloudiness and rain that did not allow the inverter to reach a value close to its nominal power of 5000 W. The

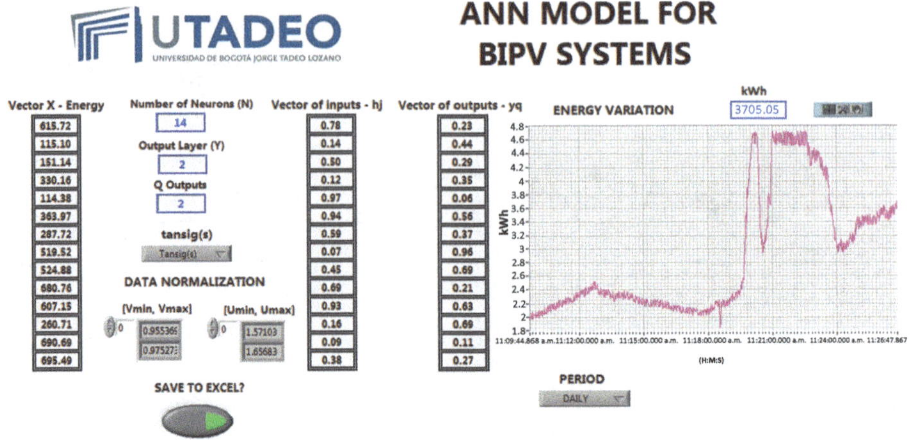

Fig. 11.4 Front panel of the virtual instrument for ANN model

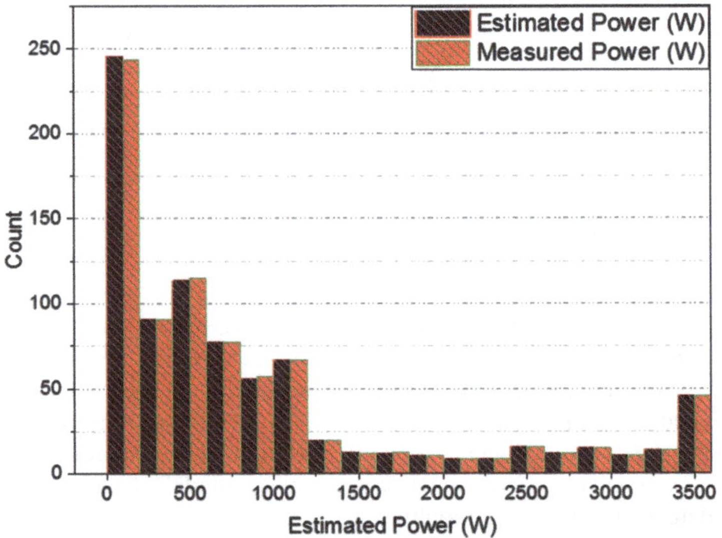

Fig. 11.5 Histogram of estimated and measured ac power of the BIPV system

maximum power generated by the inverter was 3500 W. The ac power estimated by the neural network has an approximation of 98% in comparison to the ac power measured by the monitoring system.

Also, the artificial neural network is evaluated by estimating the dc power and ac power produced by the BIPV system. The results are shown in Fig. 11.6.

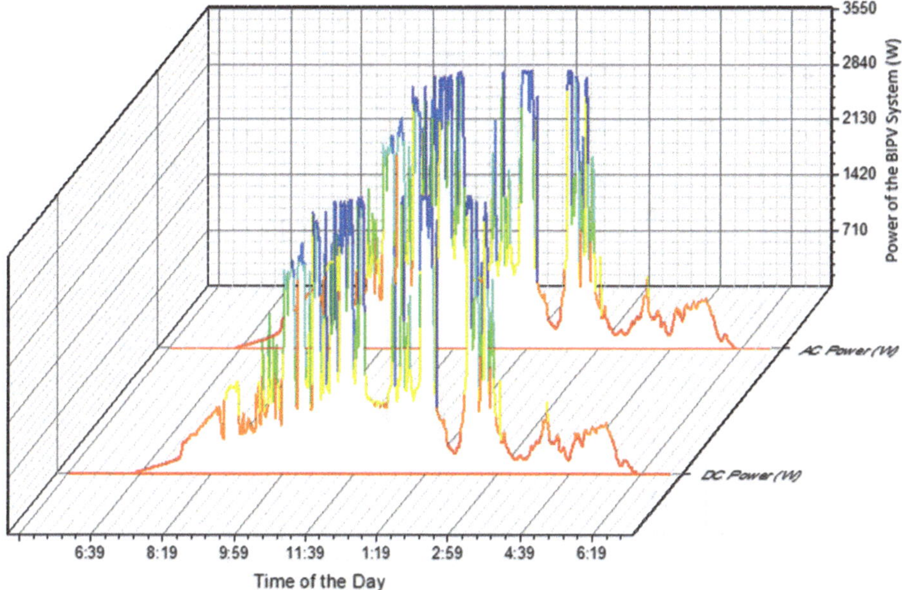

Fig. 11.6 DC and AC power estimated by the artificial neural network with four inputs for December 2nd, 2016

The area below represents the difference in power between the dc and ac power of the BIPV system. In other words, it is the inverter losses in the dc-ac power transformation process. When validating these results with those obtained by the monitoring system, there is a correlation of 98%. The inverter in this case achieved an efficiency of 92%. The 14 neurons used in this case have the capacity to estimate the result with a deviation coefficient of 0.18.

Finally, Fig. 11.7 shows both the AC energy measured by the monitoring system and the AC energy calculated with the model proposed for 2016.

The month with the highest production of AC power was May with 476.38 kWh/month and the lowest was April with 395.45 kWh/month.

The correlation of the ANN model has a coefficient of 98.6% on the AC energy of the BIPV system, between the artificial neural network and the performance data of the solar photovoltaic plant.

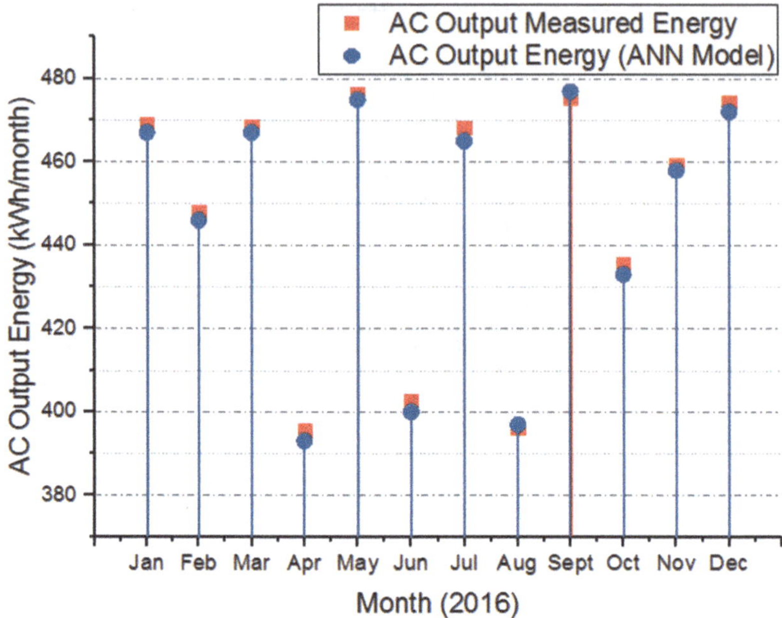

Fig. 11.7 AC output energy estimated and measured for every month of 2016

References

1. A. Campoccia, L. Dusonchet, E. Telaretti, G. Zizzo, An analysis of feed' in tariffs for solar PV in six representative countries of the European Union. Sol. Energy **107**, 530–542 (2014.) https://doi.org/10.1016/j.solener.2014.05.047
2. T. James, A. Goodrich, M. Woodhouse, R. Margolis, S. Ong, Building-integrated photovoltaics (BIPV) in the residential sector: an analysis of installed rooftop system prices. Energy **50** (2011). https://doi.org/10.2172/1029857
3. V.J. Chin, Z. Salam, K. Ishaque, Cell modelling and model parameters estimation techniques for photovoltaic simulator application: a review. Appl. Energy **154**, 500–519 (2015). https://doi.org/10.1016/j.apenergy.2015.05.035
4. V. Lo Brano, G. Ciulla, M.D. Falco, Artificial neural networks to predict the power output of a PV panel. Int. J. Photoenergy **2014**, 193083 (2014)
5. W. Zhou, H. Yang, Z. Fang, A novel model for photovoltaic array performance prediction. Appl. Energy **84**, 1187–1198 (2007). https://doi.org/10.1016/j.apenergy.2007.04.006
6. V. Sharma, S.S. Chandel, Performance and degradation analysis for long term reliability of solar photovoltaic systems: A review. Renew. Sust. Energ. Rev. **27**, 753–767 (2013). https://doi.org/10.1016/j.rser.2013.07.046
7. F.R. Chica, J. Aristizábal, Application of autoregressive model with exogenous inputs to identify and analyse patterns of solar global radiation and ambient temperature. Int J Ambient Energy **33**(4), 177–183 (2012)
8. J. Aristizábal, G. Gordillo, Performance and economic evaluation of the first grid – connected installation in Colombia over 4 years of continuous operation. Int J Sustain Energy **30**(1), 34–46 (2011)

9. T. Ikegami, T. Maezono, F. Nakanishi, Y. Yamagata, K. Ebihara, Estimation of equivalent circuit parameters of PV module and its application to optimal operation of PV system. Sol. Energy Mater. Sol. Cells **67**(1–4), 389e395 (2001)
10. A.N. Celik, N. Acikgoz, Modelling and experimental verification of the operating current of mono-crystalline photovoltaic modules using four- and five-parameter models. Appl. Energy **84**(1), 1–15 (2007)
11. J.D. Mondol, Y.G. Yohanis, B. Norton, Comparison of measured and predicted long term performance of grid a connected photovoltaic system. Energy Convers. Manag. **48**(4), 1065–1080 (2007)
12. K. Ishaque, Z. Salam, H. Taheri, Simple, fast and accurate two-diode model for photovoltaic modules. Sol. Energy Mater. Sol. Cells **95**(2), 586–594 (2011)
13. K. Ishaque, Z. Salam, H. Taheri, Syafaruddin, Modeling and simulation of photovoltaic (PV) system during partial shading based on a two-diode model. Simul. Model. Pract. Theory **19**(7), 1613–1626 (2011)
14. M.G. Villalva, J.R. Gazoli, E. Ruppert, Comprehensive approach to modeling and simulation of photovoltaic arrays. IEEE Trans. Power Electr. **24**(5–6), 1198–1208 (2009)
15. Y. Mahmoud, W. Xiao, H.H. Zeineldin, A simple approach to modeling and simulation of photovoltaic modules. IEEE Trans. Sustain. Energy **3**(1), 185–186 (2012)
16. R. Chenni, M. Makhlouf, T. Kerbache, A. Bouzid, A detailed modeling method for photovoltaic cells. Energy **32**(9), 1724–1730 (2007)
17. F. Bonanno, G. Capizzi, G. Graditi, C. Napoli, G.M. Tina, A radial basis function neural network based approach for the electrical characteristics estimation of a photovoltaic module. Appl. Energy **97**, 956–961 (2012)
18. E. Karatepe, M. Boztepe, M. Colak, Neural network based solar cell model. Energy Convers. Manag. **47**(9–10), 1159–1178 (2006)
19. K. Ishaque, Z. Salam, An improved modeling method to determine the model parameters of photovoltaic (PV) modules using differential evolution (DE). Sol. Energy **85**(9), 2349–2359 (2011)
20. M. Zagrouba, A. Sellami, M. Bouaicha, M. Ksouri, Identification of PV solar cells and modules parameters using the genetic algorithms: application to maximum power extraction. Sol. Energy **84**(5), 860–866 (2010)
21. L. Sandrolini, M. Artioli, U. Reggiani, Numerical method for the extraction of photovoltaic module double-diode model parameters through cluster analysis. Appl. Energy **87**(2), 442–451 (2010)
22. C. Huang, A. Bensoussan, M. Edesess, K.L. Tsui, Improvement in artificial neural network-based estimation of grid connected photovoltaic power output. Renew. Energy **97**, 838–848 (2016)

Chapter 12
Study and Analysis of BIPVS with RETScreen

Most countries have extensive renewable energy resources available to them, such as sun, wind, water, biomass and geothermal power. That is why renewable energy technologies (RET) are sufficiently developed and reliable to meet a larger share of the world's energy needs. In the planning and initial evaluation of a renewable technology application, it is necessary to analyze the profitability and reliability of a RET application, for decision making in the pre-feasibility and feasibility study phases of the application [1].

The Retscreen program was developed by the Natural Resources Canada (www.nrcan.gc.ca) and is supported by GEF (Global Environmental Fund), NASA and other development agencies. Retscreen International software is a tool for the study and analysis of renewable energy applications.

It is not a simulation program, but it is a key tool that helps in the process of designing clean energy applications, where it works in a comprehensive framework in the fight against climate change and pollution reduction. RETScreen International allows analyzing clean energy projects from the point of view of energy analysis, cost analysis, emissions analysis (GHG), financial, sensitivity, and risk analysis [2].

12.1 UJTL BIPVS Analysis with RETScreen

The RETScreen International software was used for the analysis of the energy system implemented at UJTL as a helping tool and support in the economic study of the BIPVS developed.

The study analysis focuses on the grid-connected Photovoltaic Generation System as a distributed generation option.

A.J. Aristizábal Cardona et al., *Building-Integrated Photovoltaic Systems (BIPVS)*,
https://doi.org/10.1007/978-3-319-71931-3_12

Fig. 12.1 RETScreen software

Site reference conditions

Select climate data location

Climate data location: Bogota/Eldorado

Show data ☒

	Unit	Climate data location	Project location
Latitude	°N	4.7	4.7
Longitude	°E	-74.1	-74.1
Elevation	m	2.546	2.546
Heating design temperature	°C	4.1	
Cooling design temperature	°C	20.8	
Earth temperature amplitude	°C	8.5	

Month	Air temperature	Relative humidity	Daily solar radiation - horizontal	Atmospheric pressure	Wind speed	Earth temperature	Heating degree-days	Cooling degree-days
	°C	%	kWh/m²/d	kPa	m/s	°C	°C-d	°C-d
January	12.9	80.6%	5.01	75.7	2.2	20.6	158	90
February	13.2	80.4%	4.66	75.7	2.2	21.4	134	90
March	13.6	81.7%	4.47	75.7	2.2	21.4	136	112
April	13.8	82.8%	3.93	75.7	2.0	21.2	126	114
May	13.8	82.6%	3.70	75.7	2.1	20.8	130	118
June	13.5	80.5%	3.79	75.8	2.5	20.2	135	105
July	13.1	78.9%	4.04	75.8	2.7	20.3	152	96
August	13.1	78.5%	4.33	75.8	2.8	21.4	152	96
September	13.1	80.0%	4.31	75.7	2.2	22.1	147	93
October	13.2	82.8%	4.33	75.7	2.0	21.4	149	99
November	13.4	84.0%	4.10	75.7	2.0	20.5	138	102
December	12.9	82.5%	4.55	75.7	2.3	20.2	158	90
Annual	13.3	81.3%	4.27	75.7	2.3	20.9	1.716	1.204
Measured at (m)					10.0	0.0		

Fig. 12.2 Environmental Data—Meteorological Location of the City of Bogotá delivered by RETScreen

12.2 BIPVS Information

With RETScreen International software help, it is possible to set (Fig. 12.1), the project location (Bogota, Colombia), project type (Power project), technology (Photovoltaic), grid type, analysis type (method 2, this method can perform cost analysis, financial analysis and emissions analysis), currency type (US dollar).

Once this data is entered, the RETScreen International program automatically loads the information from Bogota city, such as location, wind speed, radiation, humidity, temperature, pressure, etc., for an annual period, as shown in Fig. 12.2.

12.3 RETScreen Energy Model—Power Generation Project

In this page "Energy Model" is found technical parameters of the power system for the proposed case and the type of analysis (Method 2). In the evaluation of resources technical parameters are established like location and inclination of the

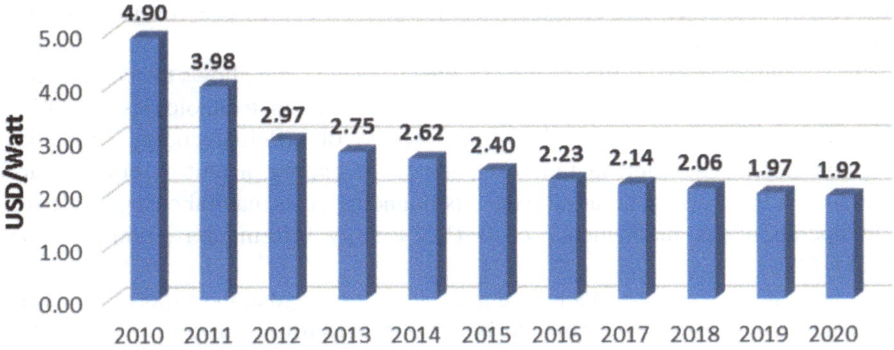

Fig. 12.3 Implementation costs projection of costs of solar photovoltaic energy systems between the years 2010 and 2020 (IRENA 2014C)

panels. Entering this data RETScreen downloads and calculates the horizontal monthly irradiation and inclined Monthly irradiation.

The energy model used in RETScreen™ has an electricity export rate that is corresponding to the projection of energy sales by the photovoltaic generation system in \$/MWh.

Advances in renewable electricity generation have been recognized as a promising trend. However, technology and innovation are driving the deployment and unblocking of new opportunities within the energy sector.

It is noted that the fall in renewable energy costs like "PV solar energy and wind energy on land" have revolutionized the industry in just a couple of years as the cost of electricity is very close to the cost of electricity generated by nonrenewable systems (International Renewable Energy Agency—IRENA, 2014b and Eclareon, 2014) [3].

Between 2009 and 2013, prices for solar photovoltaic modules fell by 65% and 70%, but their stabilization occurred in 2013 after a period of overcapacity and strong competitive pressures [3].

The following Fig. 12.3 shows the projected costs of implementing the photovoltaic solar energy system between 2010 and 2020, (IRENA, 2014c) [3].

In Fig. 12.3, the cost for implementation in 2010 was USD \$ 4.90/W, then the same analysis is reflected in 2016 where the cost of implementation is USD \$ 2.23/W, showing a decline in costs of 54.5% Compared with 2010. By 2020, the decline in the implementation would find about 60.8%. This reflects that renewable energy from photovoltaic generation is practically accessible and feasible implementation in the energy mix as a source of electricity generation. This means that they are available to individuals, cooperatives and small-scale enterprises. With recent cost reductions and innovative business models that represent the solution out of the economic network for more than 1.3 billion people worldwide lacking access to electricity [3].

According to studies carried out by the International Renewable Energy Agency (IRENA), costs of photovoltaic solar energy are between 0.11 USD/kWh (minimum base costs) and 0.35 USD/kWh (maximum base costs) for projects on a commercial scale [3, 4].

The Levelized Cost of Electricity (LCOE) is a value that allows to economically compare, in the long term, different energy generation technologies. LCOE is defined as a measure of global competitiveness of different energy generation technologies, represented in the cost per unit of energy, a unit of building and operating a life cycle generation plant, its financing costs, capital costs, fuel costs, and operation and maintenance costs (US Energy Information Administration, 2015) [5].

Based on the price of residential energy sector in Bogota, the export rate for the electricity generation system would be USD $ 0.185/KWh or USD $ 185/MWh (USD $ 1 = 2954 COP, September 2017). This value can be entered in the column of tariff of export prices for electricity in the model of energy.

The technical input data for the PV generator are selected through a library that has the RETScreen software. 24 Trina Solar panels of 245 Wp each were selected, (polycrystalline type), for a total power generation capacity of 5.88 kW, with an efficiency of 16% per panel. The inverter selected has an operating capacity of 5 kW and an efficiency of 92%. Both devices have 10% losses.

Solar electricity exported to the grid is presented in Fig. 12.4. According to this simulation, the BIPVS system has the ability to export to the grid 6.7 MWh per year, compared to the current system implemented in the CIPI building that started operation on January, 2015, which energy produced was 7.3 MWh during the first year of registration. This reflects the reliability and certainty in energy systems analysis performed under the RETScreen™ software.

In the same Fig. 12.4, the technical parameters entered and that are part of the system installed at UJTL are observed.

12.4 RETScreen Cost Analysis

In this page it is possible to find the costs associated with the application stages: system design, purchasing and procurement of equipment and materials, procurement and implementation and system's O&M system.

The overall cost of the application implemented with 6 kW was USD $ 9511, with the following characteristics:

- *Initial Costs*:
  Feasibility, development, engineering, power system and incidentals. This value is USD $ 9511 corresponding to 95.2% of the application.
- *Annual Costs*:
  They correspond to those associated with the operation and maintenance O&M cost of the system, whose annual projection was USD $ 145.

RETScreen Energy Model - Power project

Proposed case power system

Technology		Photovoltaic
Analysis type		○ Method 1 ◉ Method 2
Photovoltaic		
Resource assessment		
Solar tracking mode		Fixed
Slope	°	10.0
Azimuth	°	15.0

☑ Show data

Month	Daily solar radiation - horizontal	Daily solar radiation - tilted	Electricity export rate	Electricity exported to grid
	kWh/m²/d	kWh/m²/d	$/MWh	MWh
January	5.01	5.28	185.0	0.695
February	4.66	4.78	185.0	0.571
March	4.47	4.48	185.0	0.592
April	3.93	3.85	185.0	0.495
May	3.70	3.57	185.0	0.475
June	3.79	3.62	185.0	0.466
July	4.04	3.86	185.0	0.513
August	4.33	4.20	185.0	0.557
September	4.31	4.28	185.0	0.550
October	4.33	4.40	185.0	0.583
November	4.10	4.25	185.0	0.546
December	4.55	4.80	185.0	0.635
Annual	**4.27**	**4.28**	**185.00**	**6.678**

Annual solar radiation - horizontal	MWh/m²	1.56	
Annual solar radiation - tilted	MWh/m²	1.56	
Photovoltaic			
Type		poly-Si	
Power capacity	kW	5.88	
Manufacturer		Trina Solar	
Model		poly-Si - TSM - 245PC/PA05	24 unit(s)
Efficiency	%	16.0%	
Nominal operating cell temperature	°C	45	
Temperature coefficient	% / °C	0.40%	
Solar collector area	m²	37	
Control method		Maximum power point tracker	
Miscellaneous losses	%	10.0%	
Inverter			
Efficiency	%	92.0%	
Capacity	kW	5.0	
Miscellaneous losses	%	10.0%	
Summary			
Capacity factor	%	13.0%	
Electricity delivered to load	MWh	0.000	
Electricity exported to grid	MWh	6.678	

Fig. 12.4 Energy Model—Energy Export Tariff

- *Periodic cost*:
  It refers to the projected costs in years, where the life or replacement of parts involved in the application is determined. In this case the lifetime of inverters is 12 years. The projection of costs for the inverters in the year 12 was USD $ 1250.

12.5 RETScreen Emission Reduction Analysis

In recent years, a series of institutional actions have been developed at the national, regional and global levels focused on reducing the environmental imbalances generated by human activity and that have led to the emergence of a new market known as the Carbon Market. This market flourishes as a commitment assumed by countries, industry, the consumer market and individuals in reducing Greenhouse Gases emissions [6].

Latin America plays a very important role, offering new options for the economic development of the region, based on the implementation of projects that imply practical improvements in the production and consumption of energy, clean production and the use of renewable energies.

Colombia can be an important supplier of global environmental services by the possession of environmental assets and must recognize the great opportunity in the hands of the "Carbon Bonds." The annual GHG emission reduction potential of the total Clean Development Mechanism (CDM) projects is approximately 16.4 million tons of CO_2 equivalent, which could generate potential revenues for the country of USD 152 million [6].

12.5.1 Base Case—Electrical System (Diesel Generator)

In the analysis of GHG emissions done in RETScreen International (Fig. 12.5), the base case of the selected electrical system is Petroleum Diesel, since the building CIPI counts on a system of backup (generator set) that in case of failure of the electrical grid it would support the system load.

This case generates an emission factor of 0.933 TCO_2/MWh and a contribution of 6.2 TCO_2/year for fuel consumption of 7 MWh.

12.5.2 Proposed Case—Electrical System (Photovoltaic Generator)

Thanks to the use of photovoltaic panels, the emission factor is 0.00 TCO_2/MWh and a contribution of 0.00 TCO_2 for the generation of 7 MWh Fig. 12.5).

RETScreen Emission Reduction Analysis - Power project

Emission Analysis

- ○ Method 1
- ◉ Method 2
- ○ Method 3

Global warming potential of GHG

25 tonnes CO2 =	1 tonne CH4	(IPCC 2007)
298 tonnes CO2 =	1 tonne N2O	(IPCC 2007)

Base case electricity system (Baseline)

Fuel type	Fuel mix %	CO2 emission factor kg/GJ	CH4 emission factor kg/GJ	N2O emission factor kg/GJ	Electricity generation efficiency %	T&D losses %	GHG emission factor tCO2/MWh
Diesel (#2 oil)	100.0%	69.3	0.0019	0.0019	30.0%	10.0%	0.933
Electricity mix	100.0%	256.8	0.0070	0.0070		10.0%	0.933

☐ Baseline changes during project life — Change in GHG emission factor % -10.0%

Base case system GHG summary (Baseline)

Fuel type	Fuel mix %	CO2 emission factor kg/GJ	CH4 emission factor kg/GJ	N2O emission factor kg/GJ	Fuel consumption MWh	GHG emission factor tCO2/MWh	GHG emission tCO2
Electricity	100.0%	256.8	0.0070	0.0070	7	0.933	6.2
Total	100.0%	256.8	0.0070	0.0070	7	0.933	6.2

Proposed case system GHG summary (Power project)

Fuel type	Fuel mix %	CO2 emission factor kg/GJ	CH4 emission factor kg/GJ	N2O emission factor kg/GJ	Fuel consumption MWh	GHG emission factor tCO2/MWh	GHG emission tCO2
Solar	100.0%	0.0	0.0000	0.0000	7	0.000	0.0
Total	100.0%	0.0	0.0000	0.0000	7	0.000	0.0
				T&D losses		Total	0.0
Electricity exported to grid	MWh	7			0	0.933	0.0
						Total	0.0

GHG emission reduction summary

	Years of occurrence yr	Base case GHG emission tCO2	Proposed case GHG emission tCO2	Gross annual GHG emission reduction tCO2	GHG credits transaction fee %	Net annual GHG emission reduction tCO2
Power project	1 to -1	6.2	0.0	6.2		6.2

Net annual GHG emission reduction 6.2 tCO2 is equivalent to 2.664 Litres of gasoline not consumed

Fig. 12.5 Analysis of Greenhouse Gas Emission Reduction

GHG emission reduction summary

	Base case GHG emission tCO2	Proposed case GHG emission tCO2	Gross annual GHG emission reduction tCO2	GHG credits transaction fee %	Net annual GHG emission reduction tCO2
Power project	6.2	0.0	6.2		6.2

Net annual GHG emission reduction 6.2 tCO2 is equivalent to 2.664 Litres of gasoline not consumed

Fig. 12.6 GHG Emission Reduction Summary

12.5.3 Reduction of GHG Emissions Summary

It can be observed in Fig. 12.6 that the base case of GHG emissions is generating 6.2 TCO_2, compared to the GHG emissions of the proposed case whose contribution is 0.0 TCO_2. it is concluded that the reduction of total annual Greenhouse gases emissions with the photovoltaic system is 6.2 TCO_2, equivalent to 2664 litters of gasoline avoided.

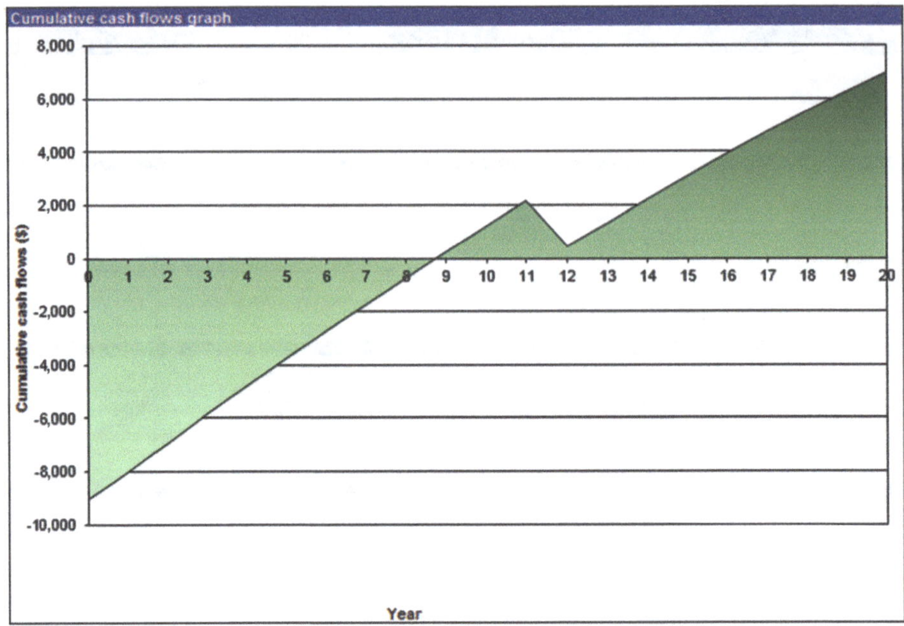

Fig. 12.7 Cumulative Cash Flow

12.6 RETScreen Financial Analysis

This page describes the important financial parameters such as the fuel staggering rate projected at 5%, the inflation rate at 6.5%, the project lifetime extrapolated to 20 years. All of these parameters are entered into the RETScreen that simulates and displays the accumulated cash flow graph (Fig. 12.7).

Figure 12.7 shows that the initial economic investment of the application is USD $ 9511 and it is recovered in a period of 8.3 years and energy savings for generation and/or sale of annual energy is 7 MWh at a cost of USD $ 0.185/KWh equivalent to USD $ 1235/year. After 9 years the generation or sale of power is reflected in net income, with a small valley in Fig. 12.7 due to the replacement of inverter. The net present value (NPV) for the application to 20 years is USD $ 6.993, with annual savings of 350 USD. This reflects that the investment for BIPVS is profitable with positive profit margins after 8 years of operation.

References

1. G.J. Leng, Section Head, Natural Resources Canada, CANMET Energy Diversification Research Laboratory (CEDRL). Lionel-Boulet Boulevard, Varennes, Quebec J3X 1S6 Canada (1615)
2. D. Thevenard, G. Leng, S. Martel, in *The RETScreen Model for Assessing Potential PV Projects*. Photovoltaic Specialists Conference, 2000. Conference Record of the Twenty-Eighth IEEE. IEEE (2000)
3. IRENA, International Renewable Energy Agency, Rethinking Energy 2014. *Towards a New Power System*. Copyright © IRENA (2014)
4. IRENA, International Renewable Energy Agency, Renewable Energy Power Cost in 2014. Copyright © IRENA (Jan 2015)
5. A. Mendoza, A. Trespalacios, *Afectación de los ingresos operacionales de empresa distribuidora de energía eléctrica por penetración de energía solar en su área de influencia*. Medellin – Colombia (2014)
6. UPME, Min Minas. *Mecanismos e Instrumentos Financieros para Proyectos de Eficiencia Energética en Colombia*. Bogotá – Colombia (2011).

Index

A.J. Aristizábal Cardona et al., *Building-Integrated Photovoltaic Systems (BIPVS)*,
https://doi.org/10.1007/978-3-319-71931-3